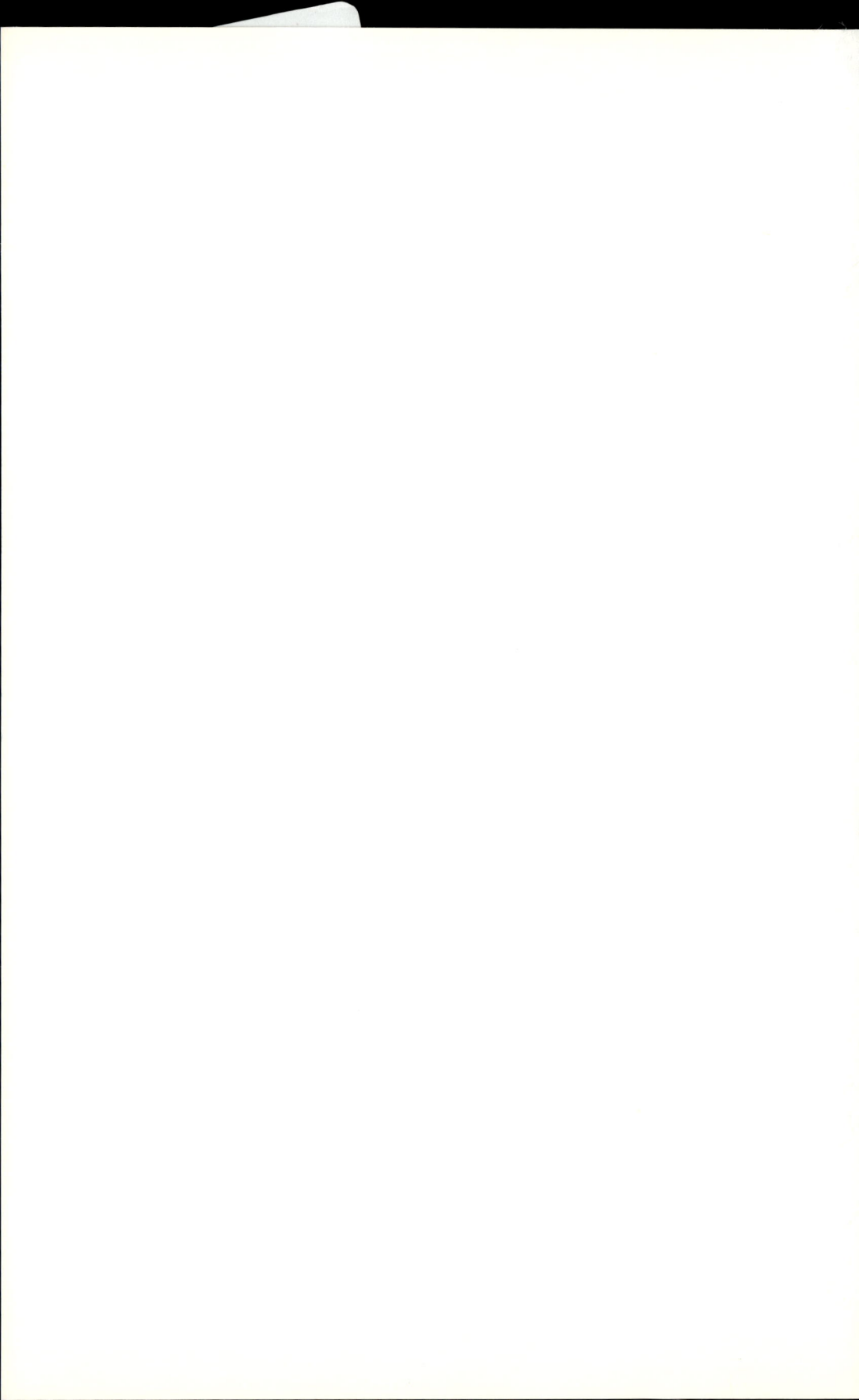

The
Cable Television
Technical Handbook

The
Cable Television
Technical Handbook

Bobby Harrell

ARTECH **AH** HOUSE

Acknowledgements

The author wishes to express his sincere appreciation to Daisy Harkness and Gail Ellington who assisted in the initial typing of this book.

A special thank you to Marilyn Holmes who worked so diligently and faithfully assisting me in the final editing of this book.

Also, my appreciation to the manufacturing companies: Scientific Atlanta, Texscan Corp., Hewlett Packard, RCA Cablevision, National Cable Television Institute, and many others who shared their knowledge and materials to make the writing of this book possible.

Again, thanks to all of you, and may God always be with each of you.

Bobby Harrell

Dedication

This book is dedicated to my mother, father, and daughter.

Preface

This book is intended for telecommunication installers, technicians, and engineers who want to improve their knowledge in the field of cable television. Inasmuch as the state of the art is changing, it is important to have good knowledge of the fundamentals of the basic cable television system. The objective of this book is to express these basic principles.

Contents

1 Introduction

1.1 What Is Community Antenna Television (CATV)?

CATV is a means by which a signal is taken out of the air and sent electronically via coaxial cable to each customer's home in the community. In the 1950s, television was still considered a technical phenomenon by a great number of people, and in many sections of the country, it was truly a phenomenon if a TV set could receive an acceptable signal. Television signals travel in a line-of-sight path; i.e., they move in a fairly straight line without adjusting themselves to obstacles. Therefore, if a mountain happens to be between the TV transmitter and the TV receiver, the result will be either no signal, or a very poor signal. Mountains and hills are not the only obstacles. Bad weather can also interfere with a TV signal. Other obstacles are man-made, e.g., tall buildings, trees, *et cetera*.

One answer to poor or weak reception was a very tall roof-top antenna. There were still several problems: one was cost; another was maintenance of the antennas; while still another was shortage of TV signals, good or bad. So, in many cases, set owners were putting up costly antennas to receive only one or two stations and were still having signal interference from weather.

Even with all the problems of television in the 1950s, it was obvious that TV was going to be a major source of entertainment for most people. Resourceful people began designing solutions to the problem of reception. Hotels, apartments, and houses began using *master antenna systems* — coaxial cable systems which carried first radio and then television signals to many sets in the same building. Because this system worked, they thought it would work for a community as well.

1.2 Community Antenna Television Becomes Cable Television

In addition to improving reception of local signals, CATV systems have come to do much more for a community. For one thing, operators began adding signals from greater distances and, using microwaves, a system owner could bring in stations from hundreds of miles away by satellite.

Cable operators found that they could carry commercial broadcast signals as well as originate local programming. The CATV system is a closed circuit television system for a whole town. Just as schools, airports, hospitals, and other institutions use closed circuit television to transmit certain "programs" within the institution, CATV systems can use their facilities to transmit programs for their subscribers only.

Most systems today are 12 to 56 channel systems. In the beginning, if an operator had five channels available and only three commercial television stations to carry, the operator naturally wanted to fill the other channels. What better way than to offer subscribers something no one else in town could get? Thus, the first programming on cable system was automated programming. The operator would purchase an automatic device which showed the time and the current temperature. Arranging this in front of an inexpensive camera, perhaps with a tape of background music playing, he "cablecast" this information to subscribers with many of the channels brought in by satellite. Other services included, for example, stock-market reports, time and weather services, alarm systems, banking at home and shop at home (see Fig. 1.1).

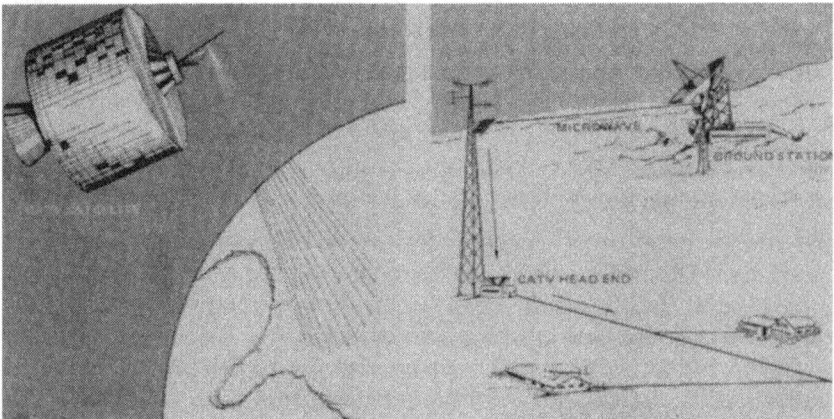

Figure 1.1 CATV Configuration
Source: National Cable Television Institute

1.3 A Combination of Wire and Wireless Communications

CATV combines the advantages of wire and wireless communications. A cable television system receives its TV and FM signals off the air, from commercial broadcast stations, where it is processed in the *head-end* and sent to the customer's home. CATV provides reception unaffected by weather or other significant interference. This is particularly important in the case of color television signals. Another important benefit of CATV is that, because programming can be originated on the system, it provides even the smallest community with an outlet for expression and advertisers can "select" their audience. For example, a program of information for mathematics students, sponsored perhaps by a university, can be shown only to the mathematics students subscribing to the service. This ensures that the sponsor will recieve the best possible audience for his program.

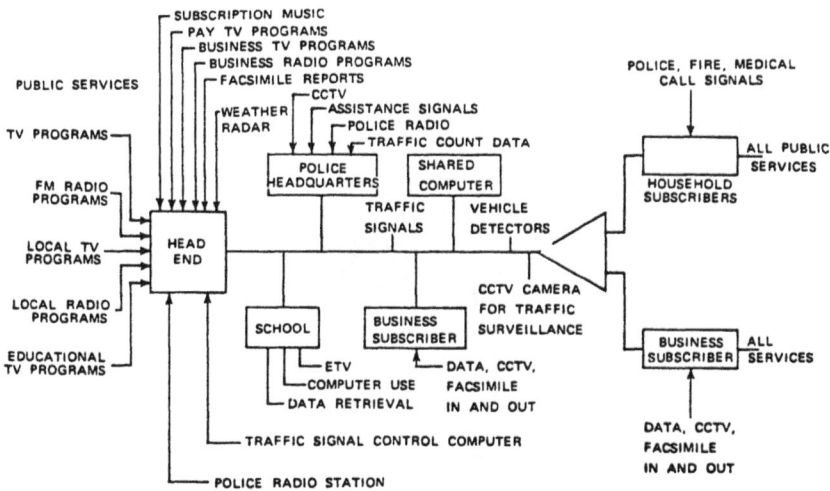

Figure 1.2 Possible Future CATV Services
Source: National Cable Television Institute

2   The Antenna Site

2.1 The Antenna System

The antenna system is normally constructed on a location that will receive good strong signal levels. A tower or pole is used to support the antenna or antennas. In most cases, the tower is located on an elevated surface such as a hill or building. The antennas are mounted in such a manner as to receive the maximum signal strength possible from the television station.

A *coaxial cable* is used to transfer the signal from the antenna to the processing equipment. In areas where the signal level is extremely low, a preamplifier is placed in the coaxial line to amplify the signal.

Great care should be exercised when using coaxial cable. Any damage to the coaxial cable will cause degradation of the signal.

The head-end building should be located as close as practical to the antenna system. This will keep the coaxial cable length short, thereby minimizing signal attenuation.

2.2 The Antenna Site

Choosing an antenna site can be a tedious process. Consideration must be given to the location of the antenna site. The ground elevation must be checked to make sure that the site is high enough to receive a good signal with a tower at reasonable height. A path survey must be performed to verify that there are no obstructions in the line-of-sight profile. If an obstruction is encountered, then the height and width of the object must be verified. Once this information is obtained, the antenna elevation can be calculated. In circumstances where the obstruction is extremely high and wide it may be necessary to select another site. After the elevation and azimuth of the proposed site is known, an interference survey is done. This survey will normally tell you the amount of interference that will be produced by neighboring radio and television transmitters. Again, if the predicted level of interference is too high, it may be necessary to choose another antenna site. After the proposed antenna site has met all the technical requirements, negotiation with the land owner must be undertaken to buy or lease the site.

2.3 Tower

A tower is used to raise the antenna or antennas to the recommended receiving elevation for good reception of the television signal. The type of tower used depends upon the size and weight of the antenna or antennas, the area of land on which it is to be located, or the structure it may be placed on or attached to, the nature and shape of the land area, and environmental conditions. Antennas are fastened to the sides of the tower, normally near the top. They can be mounted at any height on the tower depending on the best height for each television signal and the space available. No structure is to be extended above the tower. Beacon lights are placed on a tall tower to indicate a hazardous location for aircraft.

2.4 Antennas

An antenna is an electronic device used to amplify the signal received from the television station. Antennas vary widely in shape and size, but they all have one thing in common: the amplifier signal. The two most common types of antennas used are *yagi* and *log-periodic*. The yagi antenna is normally designed to receive a single television channel. The log-periodic, although similar in appearance to the yagi, is normally designed to receive several television channels simultaneously. These antennas will be discussed in greater detail later in this book.

3 The Transportation Trunk

The coaxial cable may go along a road's right-of-way, which may include a utility pole line. The coaxial cable may be attached to these same poles by agreement with the utility company through a joint or multiple use arrangement using the same right-of-way.

It is often more economical for the CATV company to build its own plant without regard to the utility. The right-of-way must then be obtained from private property owners or from a governmental entity concerned with publicly owned land.

Source: RCA Cable Construction Manual

All right-of-way should be accessible for maintenance and repair. Associated equipment is always attached to the coaxial cable and to the cable support system. If on a utility pole line, the equipment is attached to the same messenger cable as the coaxial cable. All associated coaxial cable equipment should likewise be easily accessible for maintenance, repair, and adjustment.

3.1 Cables

Each type and size of coaxial cable carries signals with its own characteristic advantages and disadvantages. The outside circumference of the coaxial cable is covered with a metallic material called a *shield*, which has good electrical and physical properties. This shield is made as uniformly as the manufacturer can maintain. The metal is either copper or an aluminum alloy. The form of this outside material depends on the rigidity required. The copper may be braided wire strands or narrow, flat strips, sometimes in two layers, or a single sheet with overlapping edges and corrugations around the circumference of the cable. However, copper is so costly that it is now seldom used. Aluminum alloys for the outer part of a coaxial cable are solid, seamless tubes of uniform dimensions.

In the center of the coaxial cable is a *conductor*, or wire. The size of the center conductor is dictated by the functions it will perform and the general characteristics of the coaxial cable. It can be made of solid copper or copper tubing, aluminum or aluminum wire with a copper-clad coating, or steel wire with a copper-clad coating.

Between the shield and the center conductor is a volume of space called the *dielectric*, or non-conductor. Something is usually put into this space to make the cable less compressible. A wide range of materials are non-conductors, from common air, inert gases, or pure water and several other liquids to a wide range of solids. Often a combination of dry air and a porous, polyethylene material is used.

An additional cover (or *jacket*) is often applied to the outside of the shield of some cable. This jacket, on the outside of the coaxial cable, is wrapped tightly against the shield, and completely encloses its circumference. Polyethylene, xelon, or forms of vinyl are used as jacket material. The thickness of the jacket will vary according to the manufacturer and type of material used. However, uniformity is still the key to quality. A stainless steel, braided armor jacket may enclose the entire coaxial cable and the first jacket. There may even be a second jacket of non-conductive material on the outside of the stainless steel braid — this is for an exceptionally hazardous environment.

The shield of a coaxial cable has multiple functions. Electrically, the shield keeps the signals within a confined space, and the regularity of the shield's shape determines in part how well the signals travel in one direction. To be worthwhile in most locations, the shield must also keep other signals out of the system and the shield is also used to carry power to associated equipment.

The ability of the cable to hold its own weight is improved with the stronger shield. The shield, if seamless or welded-seam, also serves to protect the center conductor or dielectric from contamination by water or other foreign material.

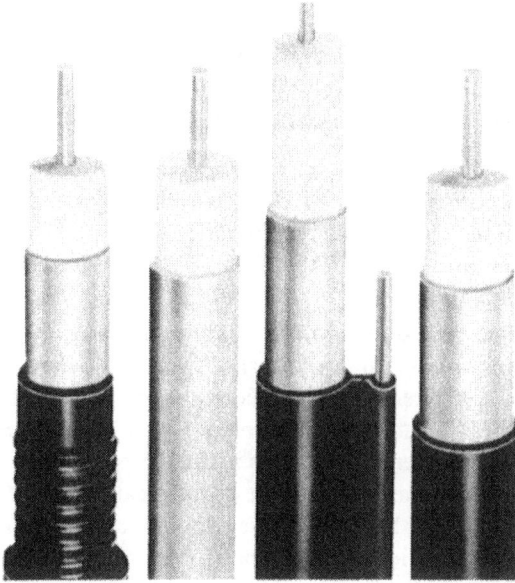

Source: Scientific Atlanta

The center conductor of the coaxial cable is multi-functional. Its outside diameter must be uniform in order to keep the signal going in only one direction. The center conductor is also used for power conduction and is usually the limiting factor in the ability of the coaxial cable to carry power. The center conductor may be a copper-coated steel strand which assists in the support of the cable in span.

The dielectric portion of the coaxial cable has a most important function in keeping the center conductor centered between the walls of the shield and keeping the two walls always parallel. The dielectric may also be bonded to the center conductor in order to help keep the center conductor from contracting more than the shield in cold weather. The most important role the dielectric plays is in maintaining uniformity, with a constant velocity, or propagation, of the signal. It should also have uniform characteristics for low loss and non-reversal of signal. The larger the diameter of the outside shield, the larger the center conductor must be to have the same impedance characteristics. The closer the dielectric is to equaling dry air or a vacuum, the better is its ability to carry signals with low loss. Therefore, the nature of the dielectric largely determines the propagation velocity and the loss characteristics.

3.2 Active Devices

Because coaxial cable is classified as a passive device and the power of the signals on the cable is dissipated or lost during transmission, this power loss must be renewed periodically along the cable. Otherwise, the signal will be of little use. Therefore, active devices are added as associated coaxial cable equipment to restore the power loss while keeping the response characteristic of the signal the same in all other aspects. The major active device used for this function is called an *amplifier*.

The amplifier may contain other active equipment that can sense changes in the power of the signal coming into the amplifier. This unit, which senses the power level changes and adjusts for them, is called an *automatic gain control* (AGC).

Because the coaxial cable does not lose power or attenuate all signals to the same degree (this is a function of the carrier wave frequency), the amplifier may have built-in characteristics which correct the different attenuations at different frequencies. In addition to the coaxial cable's varied response to the signal's frequency characteristic, changes in the temperature of the dielectric and metallic structure of the coaxial cable can also cause change in the signal. Therefore, some amplifiers have built-in sensing devices which can change the characteristic of the amplifier to compensate for the opposite characteristic in the coaxial cable. These devices are called *thermal cable equalizers.*

The design of the amplifier used on the coaxial cable determines the maximum number of amplifiers which can be used in the cascade while retaining useable signals. The number of amplifiers itself dictates what type and size of coaxial cable can be used. Therefore, *cascade-ability*, or the problems of adding amplifiers in series, is economic as well as technical. The distance between amplifiers and, therefore, the length of one continuous cable is constant in any cascade, but can vary in different cascades.

The coaxial cable is also used as a power transmission medium so that power may be put on the coaxial cable at one point to feed several active devices at other points on the same cable. The coaxial cable is connected into and out of a passive device with a third connection for power. One power source often supplies six amplifiers — three in each direction from the power source — or more.

3.3 Microwave Links

Because distance, terrain, and cost make coaxial cable impractical for use in the transportation trunk at all times, another method is used for medium and long distance. Microwave links are the common alternatives. In order to use this technology, the most practical distance between the antenna site and head-end is five miles; the maximum is several hundred miles. However, there is no technical reason that the maximum distance could not be thousands of miles. In fact, satellites are now used for sending television signals from continent to continent and could also feed signals from antenna site to head-end in a microwave-like arrangement.

Because the microwave equipment emits electromagnetic waves into the atmosphere, the transmitter and all other equipment is subject to federal licensing and inspection under the authority of the Federal Communications Commission (FCC). Therefore, the equipment must meet strict standards of performance and stability and must be operated only by those people licensed by the FCC.

Microwave transmitting equipment is usually housed at the antenna site in the same building as the processing equipment and uses the same power source. The power of the transmitter is carried to the microwave antenna by a *waveguide* that is made of copper and is in a solid, rectangular, uniform shape. Occasionally, the waveguide is a silver-plated, tough metal in a spiral shape with a flexible, air-tight outside covering, or circular in shape and pressurized with dry air or another gas. The waveguide is connected to an antenna feed which sends the radiation of microwave energy to a large parabolic-shaped reflector. The parabolic shape causes the energy to be sent in one direction with a very narrow beam, usually with less than a 3° spread.

4   Distribution System Overview

The primary purpose of the distribution system is to transport quality signals from an organization point to the subscriber. In order to accomplish this, the following signal transportation system components are required. Not included in this list is the hardware necessary to construct the system.

1. Cable
2. Trunk Amplifiers
3. Line Extender Amplifiers
4. AC Power Supplies
5. Passive (Directional Couplers and Splitters) Devices
6. Taps
7. Power Inserters
8. Connectors
9. In-Line Cable Equalizers
10. Plug-in Pads and Equalizers

The basic distribution system can be logically divided into two closely related component parts: (1) the *trunk* system and (2) the *feeder* system. These parts are sometimes rather loosely referred to as the trunk and distribution system, as in this case where the feeder system and the distribution system are considered the same. The most correct terminology, however, is *trunk and feeder systems* for the component parts and *distribution system* for the overall.

The purpose of the trunk is to transport the origination signals throughout the area to be covered by the cable system. In order to satisfy this requirement and reach the outermost system boundaries, it is sometimes necessary to cascade upwards of twenty or more trunk amplifiers. Because each amplifier contributes some amount of noise and distortion, it becomes extremely important that the system contain amplifiers that contribute minimal noise and distortion, and that the system be designed with the shortest possible trunk run length in order to minimize the number of cascaded amplifiers.

The two distortion factors which serve as cascade limiters are *noise* and *cross modulation*. Noise generally dictates the lower signal level limit and cross modulation (X-Mod), the higher. For a given cascade length, X-Mod distortion becomes objectionable at a much faster rate than does noise. As a result, X-Mod level considerations should include greater safety margins than noise level considerations.

As stated above, it is the purpose of the trunk system to transport origination signals throughout the area to be served. Once this is accomplished it is then necessary to bridge from the trunk system to the feeder system so that origination signals can be delivered to the subscriber. It is the purpose of the feeder system to accept a sample of signal energy from the trunk or transportation system and deliver this sample signal at a predetermined and consistent level to the subscriber, while at the same time providing a high degree of isolation between the trunk and the feeder.

An extremely important parameter of the distribution system is *operating level*, defined as the signal level at the output of each amplifier at the highest signal frequency carried on the system. Total system performance hinges on this single parameter, making it the most important indicator of proper system operation.

Because of the different requirements of the trunk and feeder systems, different operating level considerations are necessary. The trunk system has by far the greatest number of amplifiers in cascade, and as a consequence must rely mainly on cascade considerations for setting operating levels. Amplifier cascade theory, when considered with respect to desired carrier-to-noise ratio and amplifier overload, dictates an optimum trunk amplifier gain of roughly 22 dB. Because the entire system is operated on the basis of unity gain (i.e., the total signal loss in front of an amplifier is exactly equal to the gain of the amplifier so that the net gain is unity) the optimum trunk amplifier spacing is, therefore, 22 dB.

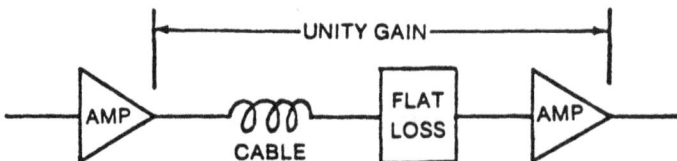

Figure 4.1 Cableloss + Flatloss = Amplifier Gain
Source: Scientific Atlanta

Because the trunk contains a larger quantity of amplifiers in cascade and operates at a lower level than the feeder, it has the most predominant influence on the overall system carrier-to-noise ratio. In fact, the feeder will generally influence the overall system carrier-to-noise ratio by only about 1 dB.

The operational level of the feeder is based mainly on efficiency. Within the feeder system, the idea is to distribute signals to as many subscribers as possible per amplifier and associated length of cable. Therefore, the higher the subscriber (*tap*) count for a given amplifier and length of cable, the higher the feeder efficiency. This concept would indicate that feeder efficiency would be closely aligned with high operating level, and indeed this is generally the case. Because the signal suffers loss as it travels down the cable, and also suffers loss as it passes through each tap, the number of taps that can be installed in a given feeder line becomes a function of feeder operating level. There is, however, an upper limit to the level at which the feeder can be operated. Since X-Mod is a result of overload, and overload a function of high operating level, the upper level limit of the feeder is set by the degree of X-Mod distortion considered acceptable.

In general, the feeder system will have a maximum cascade of two amplifiers. This seems to be a small number of cascaded amplifiers. However, because each feeder leg can contain as many as three line extender amplifiers, the subscriber area covered by the two-amplifier cascade becomes considerable. (See Figs. 4.2 and 4.3 for a typical system layout.)

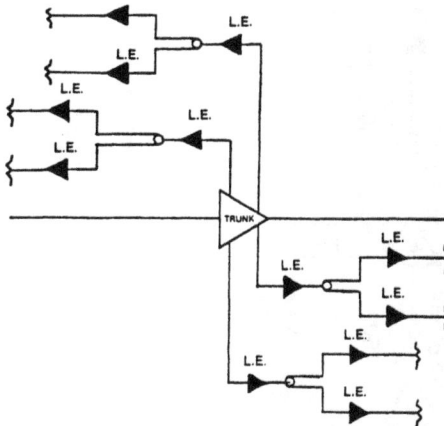

Figure 4.2 Feeder Containing a Small Quantity of Amplifiers in Cascade Operated at a Much Higher Level than the Trunk. Note: Because of this higher level the feeder will have the most predominant influence in an overall system X-Mod distortion for a given change in system operating level.

Source: Scientific Atlanta

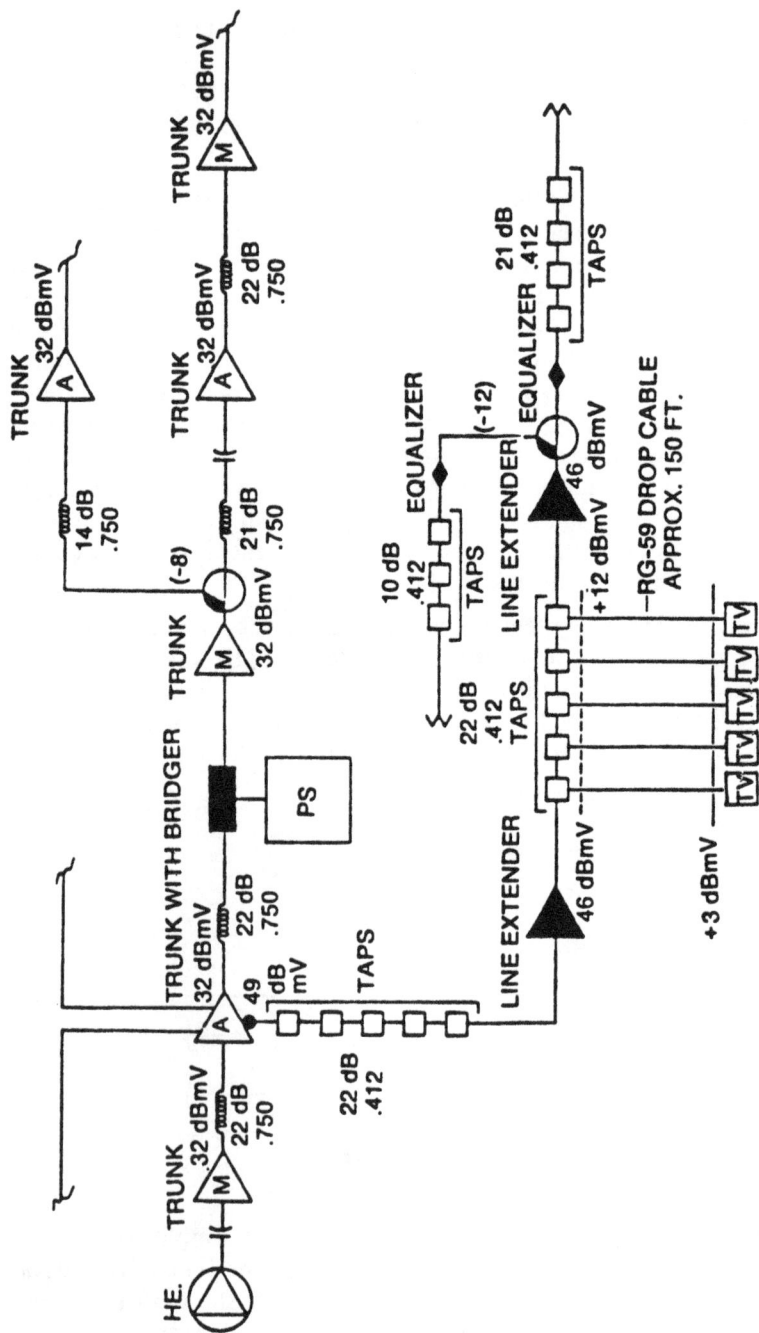

Figure 4.3 Distribution System
Source: Scientific Atlanta

5 Coaxial Cable and Equalization

Coaxial cable is the most frequently used medium for signal transportation in existing cable systems. It is through the coaxial cable medium that signal energy is transported from an organization point to the input of the subscriber's TV set. If there were no loss on coaxial cable, signal energy could be transported infinite distances without the need of amplifiers. In reality, however, a particular cable at a particular temperature has an attenuation characteristic that is directly related to frequency. Furthermore, as the temperature changes, the frequency attenuation characteristic of the cable also changes. It is, therefore, the frequency temperature attenuation characteristic of the cable that dictates the need for equalized amplifiers and initial system balance and alignment. To fully understand these requirements, it is necessary to examine the frequency attenuation characteristics of cable and the temperature related changes that occur.

5.1 Coaxial Cable

Coaxial cable is constructed of two conductors which have a *common axis* (co-axis, or coaxial). The common axis is achieved by arranging the two conductors as concentric circles, as shown in Fig. 5.1. Conductor number 1, the center conductor, and conductor number 2, the shield, are held uniformly separated throughout the length of cable by an insulating or dielectric material. The equivalent electrical representation of a length of coaxial cable is shown in Figs. 5.2 and 5.3.

In Fig. 5.2, a lossless length of coaxial cable is represented as a series connection of inductors for both the shield and center conductor, and a shunt connection of capacitors between the shield and center conductor. Because the two conductors of the cable carry an alternating signal current, each exhibits the property of inductance. Also, because the cable is essentially two conductors separated by a dielectric, it effectively exhibits the characteristic property of capacitance.

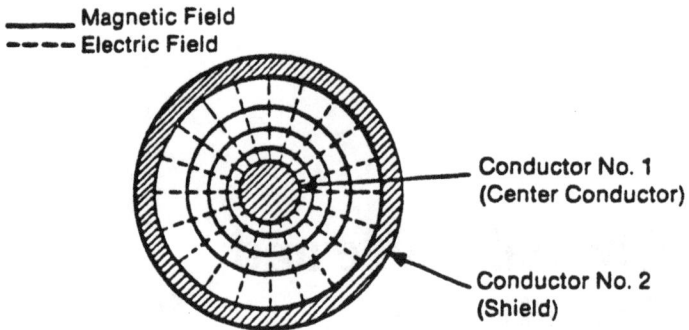

Figure 5.1 Coaxial Cable
Source: Scientific Atlanta

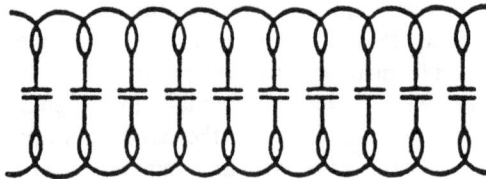

Figure 5.2 Schematic Representation of the Distributed Inductance and Capacitance of a Transmission Line
Source: Scientific Atlanta

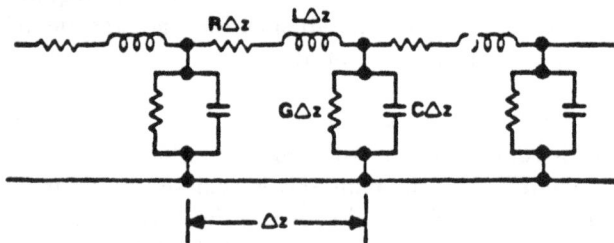

Figure 5.3 Approximate Representation of a Short Section of Transmission Line
Source: Scientific Atlanta

Figure 5.3 shows a more realistic representation of a typical length of coaxial cable. In addition to inductance and capacitance, a typical cable also exhibits series resistance associated with each conductor and shunt conductance associated with the dielectric. Both conductor resistance and dielectric conductance demonstrate frequency related loss and are the two properties directly responsible for the variation in cable attenuation with frequency.

Each of the four properties above is uniformly distributed throughout the length of cable and measured as resistance, conductance, capacitance, and inductance per unit length of cable.

For the sake of completeness, it is worthwhile at this point to very briefly summarize the mathematical relationships that govern the characteristics of coaxial cable. Because the derivations of these expressions are complex and involve the use of calculus, only the resultant expressions are given.

Examination of Fig. 5.3 reveals that the ac series impedance Z can be written as

$$Z = R + j\omega l \tag{5-1}$$

where

R = resistance in ohms

L = inductance in henries

$\omega = 2\pi F$

F = frequency in hertz

and the ac shunt admittance Y can be written as

$$Y = G + j\omega c \tag{5-2}$$

where

G = conductance in mhos

C = capacitance in farads

$\omega = 2\pi F$

F = frequency in hertz

Relating ac impedance (Z) and ac admittance (Y) as a ratio and taking the square root of the result produces an important additional property called characteristic impedance Z_o. The symbolic expression for characteristic impedance is shown below:

$$Z_o = \sqrt{\frac{Z}{Y}} = \sqrt{\frac{R+j\omega l}{G+j\omega c}} \tag{5-3}$$

Again relating ac impedance (Z) and ac admittance (Y), but now as a product and taking the square root of the result, produces another equally important property called the *propagation constant*. The propagation constant is of particular importance here because it provides a means to predict the variation of signal voltage and current as a function of cable length. The propagation constant (gamma) is expressed symbolically as shown below:

$$\gamma = \sqrt{YZ} = \sqrt{(R + j\omega l)(G + j\omega c)} \tag{5-4}$$

In general, the propagation constant is a complex quantity having both a real and an imaginary part. The real part α describes signal attenuation per unit length of cable and the imaginary part β describes the variation in phase position (indirectly, the velocity of propagation) per unit length.

$$\gamma = \alpha + j\beta \tag{5-5}$$

where

α = attenuation constant (neper per mile)
β = phase constant (radians per mile)
1 neper = 8.686 dB

The above relationships are those required to totally characterize a particular cable. An important point to note is that cable attenuation is a function of conductor resistance and dielectric conductance, and not simply a result of capacitive and inductive reactance. To carry this a step further, it can be shown that the zero frequency limit expressed by the attenuation constant is

$$\gamma = RG \tag{5-6}$$

Because this is a real quantity, it represents only the attenuation constant and the phase constant for this limit is zero.

As the frequency increases, the propagation constant also approaches an upper limit. The upper frequency limit of the propagation constant is found to be

$$\gamma = \left[\frac{R}{2} \sqrt{\frac{C}{L}} + \frac{G}{2} \sqrt{\frac{L}{C}} + j\omega \sqrt{LC} \right] \tag{5-7}$$

The real part

$$\alpha = \frac{R}{2} \sqrt{\frac{C}{L}} + \frac{G}{2} \sqrt{\frac{L}{C}} \tag{5-8}$$

represents the attenuation constant with

$$\frac{R}{2} \sqrt{\frac{C}{L}}$$

being the attenuation caused by energy losses in the conductors, and

$$\frac{G}{2} \sqrt{\frac{L}{C}}$$

being the attenuation caused by energy losses in the dielectric. These last relationships demonstrate the mechanism by which frequency-related cable attenuation actually occurs. Because the high frequency attenuation constant is directly related to the sum of the variable quantities of conductor resistance (R) and dielectric conductance (G) and as these two loss factors increase with frequency, the cable loss also increases with frequency.

The foregoing relationships predict that coaxial cable has an attenuation characteristic that is a function of frequency. Furthermore, the cable attenuation characteristic is found to vary with frequency in an exponential manner. The basic cable rule-of-thumb which describes this relationship states: *Attenuation in cable caused by a variation in frequency will follow the square root of the ratio of the frequencies.* This relationship is expressed symbolically as shown below:

$$\frac{A_H}{A_L} = \sqrt{\frac{F_H}{F_L}} \quad \text{or} \quad \frac{A_L}{A_H} = \sqrt{\frac{F_L}{F_H}} \tag{5-9}$$

where

A_H = attenuation at highest frequency (F_H) in dB

A_L = attenuation at lowest frequency (F_L) in dB

F_H = highest frequency (MHz)

F_L = lowest frequency (MHz)

As an example of the use of the cable loss relationship, suppose that a cable manufacturer's data sheet specifies a 300 MHz cable loss of 0.90 dB/100 ft at 68° cable loss at 50 MHz is desired, the rule-of-thumb expression can be applied as follows:

$$\frac{A_L}{A_H} = \sqrt{\frac{F_L}{F_H}} \tag{5-10}$$

where

A_H = 0.90 dB/100'

F_L = 50 MHz

F_H = 300 MHz

and

$$\frac{A_L}{F_L} = A_H \sqrt{\frac{F_L}{F_H}} = 0.90 \text{ dB}/100' \sqrt{\frac{50 \text{ MHz}}{300 \text{ MHz}}}$$

$$= 0.40 \text{ dB}/100' \text{ at } 68° \text{ F at } 50 \text{ MHz} \tag{5-11}$$

Within the CATV industry it is accepted practice to specify cable length in dB of attenuation at the highest frequency of operation. This practice establishes a common reference and ensures industry-wide verbal and written consistency in specifying that cable length dB of attenuation at the highest frequency defines the maximum cable loss, but does not indicate the exponential, or nonlinear, loss exhibited at lower frequencies. Figure 5.4 shows the frequency *versus* attenuation characteristic of a 20 dB length of .750 cable, and Fig. 5.5 shows the same characteristic for .500 cable. From these figures it is obvious that cable exhibits a different loss factor at each channel.

Also of interest, but not so obvious, is that the same mid-frequency channels exhibit a loss factor which is different for the two sizes of cable. The frequency *versus* attenuation characteristic of cable demonstrated in Figs. 5.4 and 5.5 is referred to most often as tilt, true tilt, or slope.

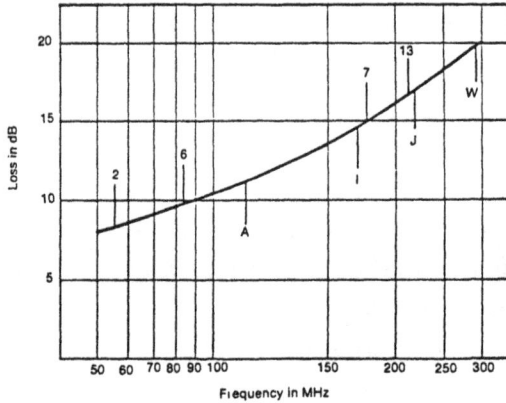

Figure 5.4 .750 Cable Loss for 20 dB Spacing/2222.2 ft./Shape Factor 0.10
Source: Scientific Atlanta

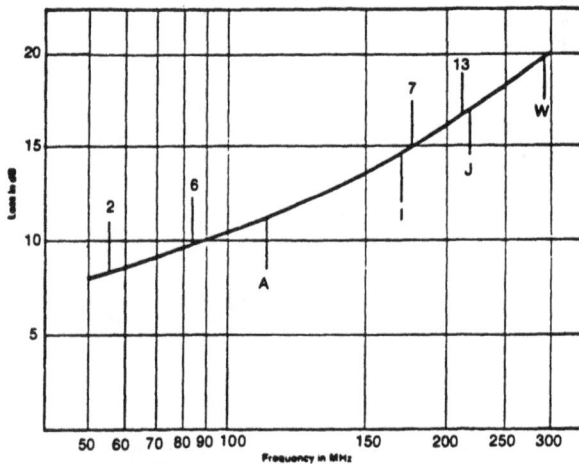

Figure 5.5 .500 Cable Loss for 20 dB Spacing/1525.7 ft./Shape Factor 0.09
Source: Scientific Atlanta

5.2 Equalization

The statement was made earlier that if it was not for cable loss there would be no need for amplifiers. This, of course, assumes only cable throughout the system and no flat loss due to taps, passive devices, and mechanical components. The implication of this statement is that there should be unity gain at all frequencies throughout the system; a fact which is indeed true and is the basis for both system design and operation.

However, in order to achieve unity gain at all operating frequencies, it is necessary to somehow compensate for the frequency attenuation (*tilt*) characteristics of cable. This could be accomplished through the design of an amplifier that would exhibit a gain characteristic exactly the inverse of cable tilt. Another method would be to design an equalizer network which, when placed in series with a given length of cable, would further attenuate all lower frequency components so that the loss at all frequencies is equal. A flat gain amplifier could then be placed at the output of the equalizer network in order to restore original signal levels. These concepts are illustrated in Figs. 5.6 through 5.10.

Both of the above concepts represent valid methods to compensate for cable loss and achieve unity gain. In fact, most modern equipment uses one or the other of these methods or, more commonly, a combination. It becomes a requirement, therefore, that system designers and technicians be able to determine the value of cable equalizers in order to ensure unity gain in both system design and operation.

Cable equalizers are generally available in predetermined cable equivalent increments, based on the requirements of a particular manufacturer's equipment. Although each manufacturer's incremental steps may differ in size and range, the cable equivalent values all have exactly the same meaning. The term *cable equivalent* means simply that a 20 dB, 300 MHz equalizer (any manufacturer) is inversely equivalent to, and will equalize, a 20 dB, 300 MHz length of cable. As an example, if the above 20 dB length of cable is connected in series with a 20 dB equalizer (Fig. 5.9), the resulting frequency response for the series combination will be flat.

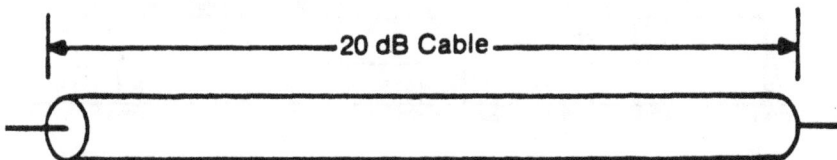

Figure 5.6 20 dB Cable
Source: Scientific Atlanta

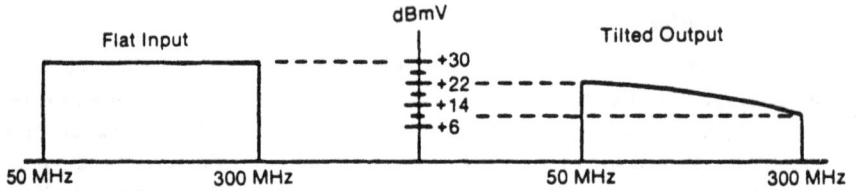

Figure 5.7 Cable Attenuation
Source: Scientific Atlanta

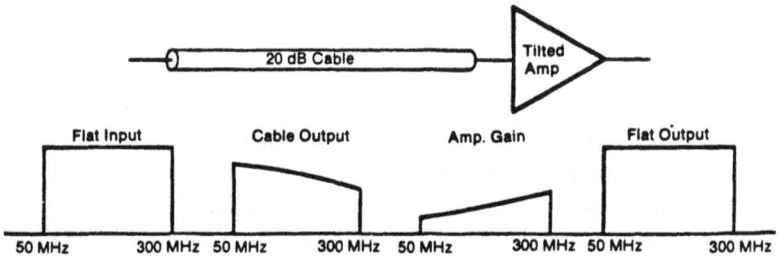

Figure 5.8 Cable and Tilted-Gain Amplifier
Source: Scientific Atlanta

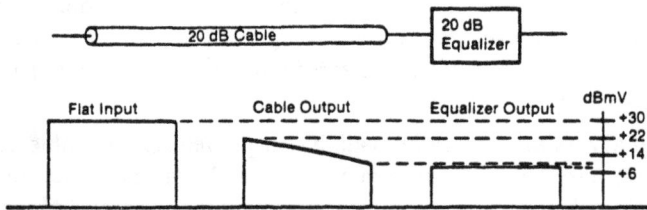

Figure 5.9 Cable and Equalizer
Source: Scientific Atlanta

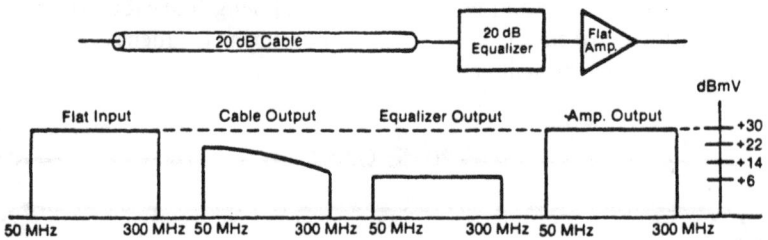

Figure 5.10 Cable, Equalizer, and Flat-Gain Amplifier
Source: Scientific Atlanta

The series combination will, however, exhibit a 20 dB flat loss as a result of the equalized 20 dB of cable and an approximate 1 dB (or less) additional flat loss due to the inherent losses within the equalizer.

The preceding suggests that the value of the equalizer for a particular amplifier location is directly related to the dB cable length in front of the amplifier. This conclusion is true with a single qualification: if the amplifier contains built-in equalization, as most do, the cable equivalent of the built-in equalization must be subtracted from the dB cable length in order to determine the correct equalizer value. Figure 5.11 (examples 1 and 2) will illustrate these points.

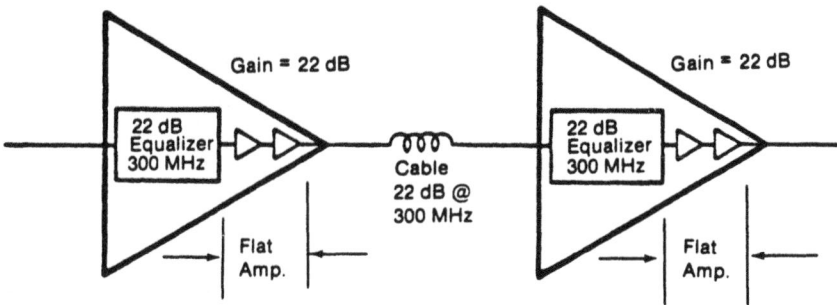

Figure 5.11 Example 1: No Built-In Amplifier Equalization
Source: Scientific Atlanta

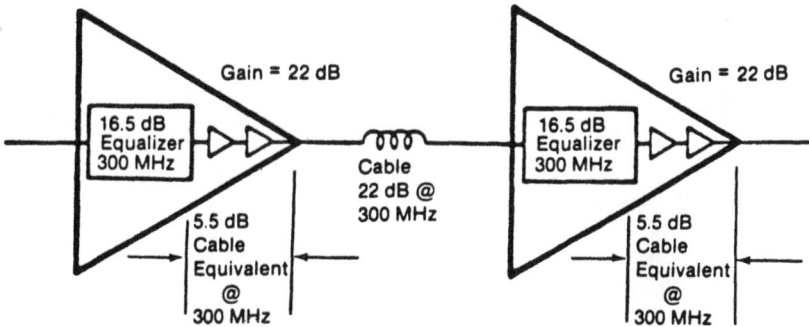

Figure 5.11 Example 2: Built-In Amplifier Equalization
Source: Scientific Atlanta

6　The Distribution System

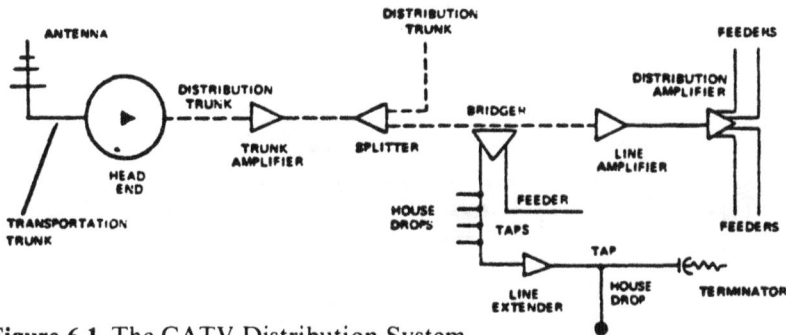

Figure 6.1 The CATV Distribution System

The distribution system is the most important part of the system. When the signals leave the head-end, they enter the portion of the system known as the *distribution trunk*. This portion is similar to the transportation trunk, but there are important differences. From the distribution trunk, the signals enter the *feeder system*. This system is also similar to the transportation trunk, but again with significant differences. The final step in the transfer of signals is from the feeder system to the *subscriber drop*. This is the part of the system that is probably the most important to the CATV system as a whole. No matter how good the previous portions of the system are, problems in the subscriber drop portion seem to cause the greatest loss of customer revenue.

6.1　The Distribution Trunk

The cable leaving the head-end, going to a place of high population concentration, is relatively large and extremely efficient in transporting the signal. Such highly efficient cable is used because at the point of consumption the signal must be split into many parts to enable the various customers concentrated in this area to have a sufficient portion of signal to operate their television receiver. This cable is called the distribution trunk.

The details of the distribution trunk system are approximately the same as the details discussed previously in the transportation trunk system. The cables are large, and the amplifiers are spaced a great distance apart.

A high degree of grounding occurs in the distribution trunk because there is a greater possibility of human contact with the equipment in more heavily populated areas. Also, the extensive use of power lines by the utility company, street lighting, and other electrical hazards in these more populated areas necessitates more grounding of the system, both for protection of the personnel working on the system and the protection of the system itself.

The feeder portion of the system introduced the great number of cables that make up the primary part of the system. From one distribution trunk there may be almost any number of feeder cables, and from each feeder cable there may be almost any number of subscriber drops.

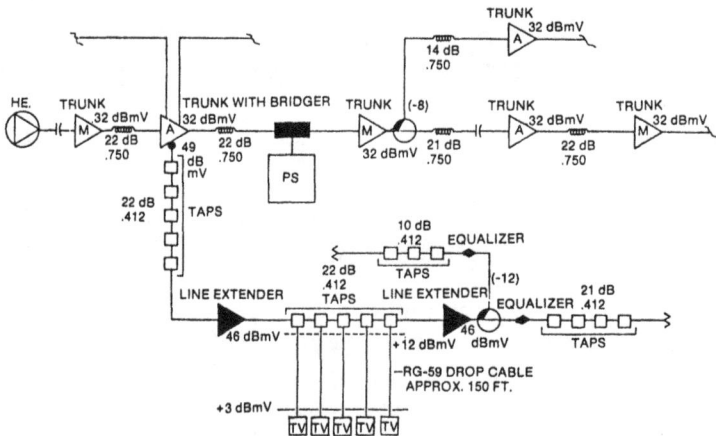

Figure 6.2 Distribution System: Trunk and Feeder Example
Source: Scientific Atlanta

Although there are amplifiers used in this portion of the system, their use is kept at the minimal level. They may be classified as *distribution amplifiers* (bridger amplifiers), which accept the signal from the distribution trunk and divide it into a number of signals for the feeder cable. The next classification of amplifiers is *extension amplifiers*. These amplifiers are used when a new portion of the system is connected to an old portion of the system in order to extend the feeder cable beyond the point of its natural ability to conduct. The third amplifier classification is *active taps*. These active taps are used to remove the signal from the feeder and send it to subscriber drop cables which are longer than normal. These long drops necessitate some type of amplifier to reduce loss, so that the signal at the subscriber's set may be kept at a level high enough to be useable.

6.2 The Subscriber Drop

This portion of the system is the final link in the chain that begins at the antenna site, goes through the transportation trunk into the head-end, out into the distribution trunk through the feeder system and, finally, into the subscriber drop. The weakest link in this chain is the last one, the subscriber drop. This is where most of the work is carried on and where most of the handling of the equipment occurs. There are constant connections, reconnections, disconnections, and other changes in this portion of the system. Therefore, the greatest number of problems occurs in the subscriber drop. The greatest effect on the quality of the signal usually occurs in the subscriber drop. The greatest number of complaints are usually answerable by correcting some problem in the subscriber drop. A new technician almost always begins his or her work in this portion of the system. Therefore, it is imperative that the technician recognize the importance of this portion of the system and do a good job, because this is where the most difficulty can occur.

6.3 Coaxial Cable

The coaxial cable used in the subscriber drop is much the same as the cable used in the rest of the system. Cable is made up of jacket (sometimes with integral messenger molded in), shielding, dielectric (usually foam), and a center conductor. Where differences occur, they are usually in the shielding and size of the cable. This portion of the system generally uses cable approximately one-quarter of an inch in diameter. The shielding of this cable is either braided copper wire, braided aluminum wire, foil, or, in some rare cases, lightweight aluminum tubing. Although the braid was the most common shielding used in the past, it is being replaced in most applications by a lightweight foil shielding, because the braid, by virtue of its construction, has holes throughout where the wires cross one another. On the other hand, the foil has complete continuity, because there is no place for radio frequency interference to enter the cable and, therefore, cannot affect the signal.

Various types of connectors, both cable-to-cable and cable-to-equipment, are used for connecting the drop cable to the tap-off equipment and for splicing to other pieces of equipment. Taps obviously are significant, because these are the connections between the subscriber drop and the feeder. Splitters are used inside the building if there is more than one set to be connected to the drop.

These splitters are generally two-way, passive splitters and, therefore, attenuate the signal to some degree. Thus, there is a limit to how many splitters may be used in one drop without further active amplification.

The other, more important passive device in the subscriber drop is the *transformer*, which matches the impedance of the cable to the impedance of the television set. Impedance is simply a way of saying that the electrical characteristics of the cable do not match the electrical characteristics of the antenna connections of the television set. Therefore, there must be some device for transforming the signal from the cable to the set. Because the CATV system is designed to carry the signal at a certain impedance, the transformer is always located immediately on the back of the consumer's television receiver. Thus, the signal is maintained at the precise level and characteristic necessary for good distribution up to the very last point possible for insertion into the receiver. This eliminates chances for interference and helps control other problems.

Further passive devices are used in the grounding equipment for the subscriber drop system. These passive grounding devices are extremely important, because the television set is handled constantly by people in the home. If grounding is not properly constructed, a lightning storm or other electrical hazard could cause damage to the set, the home, or the subscriber himself. Therefore, proper grounding of the installation itself must be considered of prime importance.

The grounding of the drop cable generally occurs at the subscriber's dwelling. This is because the most logical place to ground this portion of the cable is at the building attachment which supports the cable span (i.e., the length of cable from the tap to the building).

Greatest amount of diversity is found in the subscriber drop portion of the system. Thus, a technician who is familiar with these variations in the subscriber drop portion of the system, and who can handle a great variety of special installation problems, is now, and will remain, one of the most important people in the system. It has often been said that if the installation is good, the technician will not be required. That is, if the installation is done properly, all ratio frequency interference is properly eliminated by making good connections, water is eliminated by proper water-proofing of these connections, and equipment is not mishandled during the connection procedure, then technician will not be required to troubleshoot this portion of the system. This especially is true since most all devices are passive and do not require significant service.

The increased use of satellite communications may cause a significant change in cable television in the future. Rather than using the antenna site, trunk transportation system, microwave techniques, and other means of gathering and transporting the signals, the satellite might take over all of these functions and beam to some central head-end within the community. From this point, after minimal processing, the signals are sent out over a standard distribution system to the various subscribers in the community. Thus, the cable television system, which is now a rather isolated, local phenomenon, may become a genuinely national or international phenomenon, using satellites for inter-connection between various localities. This will become more significant when two-way communication, using CATV techniques, occurs.

It is possible that in the future the communication between people, which now depend primarily upon audio communication and the telephone, may depend heavily upon visual communication using the television set and in conjunc-tion with the telephone service. Thus, CATV may be the communications utility of the future, rather than a means of transferring entertainment and information from one point to another.

Source: Warner-Amex

The ramifications of satellite communications and a great number of chan-nels available to the subscriber (in some estimates as many as 200 channels) indicate that the future CATV systems may carry a great deal of information other than entertainment or standard communications. It is quite possible that the CATV system could become the utility meter reader i.e., the meters for electricity, water, and other services in the home could be connected directly to the coaxial cable, and the readings could be taken at some distant point. In addition, with two-way communication techniques, the subscriber could remain at home and do his or her shopping using his television set. Thus, the individual merchants within a community could have a hand-held, portable camera and sound installation. The consumer would request that the merchant show him the particular item of interest, whether it be food, clothing, *et cetera*, and he could make his selection based upon the image on his television set and what the merchant tells him about the item. This is quite similar to the current practice of using catalogues and telephones to place an order with any of the large mail-order merchants. Obviously, such techniques could either eliminate the need of catalogues or supplement them considerably.

7   Ohm's Law

Ohm's law can be most simply defined by saying that, given three electrical quantities (voltage, current, and resistance), current flow is directly proportional to total circuit voltage and inversely proportional to total circuit resistance.

This is probably the most important single theorem in the field of electronics simply because it is so useful. It is a straightforward and easily understood law, but at the same time it forms the basis for highly complex theory, and circuit and system analysis.

7.1 An Explanation of Ohm's Law

We know that Ohm's law considers three quantities: (1) voltage or electrical pressure represented by E and measured in volts, (2) current flow represented by I and measured in amperes, and (3) resistance to the flow of current represented by R and measured in ohms. The law states that voltage equals current times resistance ($E = IR$).

A hydraulic system can be used as an analogy. Let electrical pressure be the same as hydraulic pressure, current flow be the same as the flow of hydraulic fluid, and electrical resistance be the same as the degree of roughness on the interior of a pipe in the hydraulic system.

Just as a pump provides hydraulic pressure in a hydraulic system, so does a battery provide electro-motive force (voltage) in an electrical system. As the fluid flows in a hyraulic system, so does current (the movement of electrical charges) flow in an electrical system. Resistance in a hydraulic system, such as roughness of a pipe or a sudden change in pipe diameter, is comparable to the resistance to current flow in a piece of wire or in a specific, compact, electrical component called a resistor. For a fixed flow of hydraulic fluid, the higher the degree of roughness (resistance) in the pipe, the harder it will be to make the fluid move through the pipe and, therefore, the greater the pressure must be. Likewise, for a fixed flow of electrical current, the higher the electrical resistance, the greater the voltage must be. Also, for a fixed hydraulic pressure, the greater the pipe resistance, the lower the flow of hydraulic fluid, because the fluid will not move as easily. Likewise, for a fixed voltage, the greater the electrical resistance, the lower the flow of current.

7.2 Ohm's Law Applied to Power

A law concerning power is derived directly from Ohm's basic law. If a certain level of energy is supplied to an energy absorbing device, then at least part of this energy will be absorbed and dissipated in some fashion, or in some cases stored for later use. Accordingly, the amount of energy stored for use elsewhere will be less than the original amount. In electricity, this energy is power measured in watts. From Ohm's law ($E = IR$), we see that, for a fixed current flow, if we have a specific fixed point of resistance, there will be a pressure or voltage drop across this point of resistance. This amount of energy, or power absorbed by the resistance, is equal to the flow of current times the pressure or voltage drop which occurs. In other words, power absorbed by a resistance equals the amount of current times the quantity of voltage drop across the resistance ($P = IE$).

Similarly, the power absorbed in an entire system is simply the total voltage or total pressure in the system times the current flow through the system. Power required in a system equals the total current through the system times the total voltage across the system.

7.3 Ohm's Law and CATV Systems

In order to apply the tenets of Ohm's law to a specific area in CATV, observe:

1. Every item of electrical or electric equipment requires a certain amount of power (P) for operation. Therefore, when designed for a specific fixed voltage, the item of equipment will absorb the required amount of power by drawing sufficient current as needed ($P - IE$).

2. All electrical conductors (wire and cable) have a specific, very small value of dc resistance, in ohm's per foot, which is inversely proportional to the diameter of the conductor.

From the second observation given Ohm's law ($E = IR$), any section of wire carrying current will have an extremely small voltage drop per foot. As current increases, the voltage drop per foot increases. For a real life situation, the voltage drop over a very long length of wire (thousands of feet) can become appreciable.

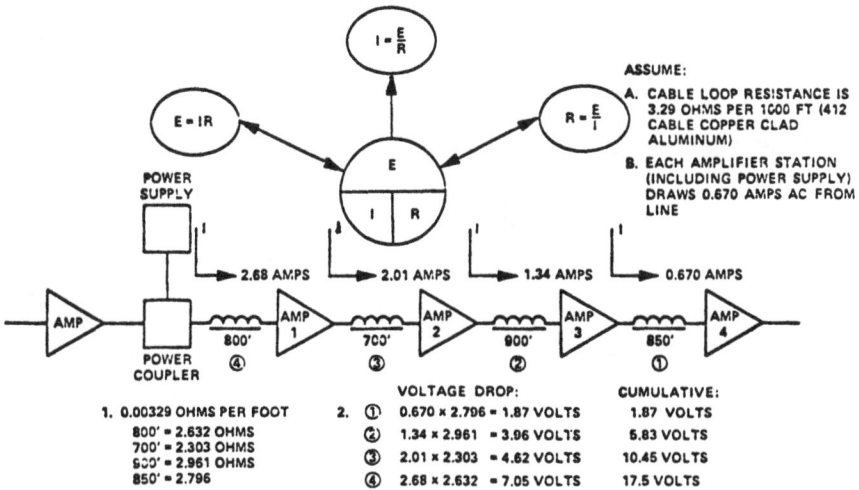

Figure 7.1 Ohms Law

Power companies alleviate this problem, which occurs over the long distances between main power stations and sub-stations, by using very high voltages. It follows (from the first observation) that a great deal of power can be delivered using a low value of current which keeps the voltage drop between main station and sub-station small.

In designing CATV equipment, manufacturers have employed this same principle. Previously, all CATV products were designed to use 30 volts, but products are now designed to use either 30 or 60 volts. Because a single power supply can provide only a limited amount of power, by using 60 volt power supplies, more equipment can be driven by a single power supply. This is because the voltage drop (and, therefore, power loss) over a specific distance due to cable resistance will be less (because less current is flowing through the cable).

The end result will be fewer power supplies required throughout the CATV system, less total system power consumption, and a better feeder to trunk ratio, which consequently provides a sizeable reduction in system cost.

It should be noted that because of the high power requirements of two-way systems, it is nearly mandatory to use the higher voltage (60 volt) method for efficient operation.

7.4 dB and dBmV

In CATV systems, technicians are constantly confronted with measuring relationships between various signal levels and measuring exact signal levels throughout the entire system. The calculations for determining these relationships and exact values should be as simple as possible in order to decrease the possibility of error. To accomplish this, a mathematical system has been devised based on the decibel (dB).

7.5 dB Applied to CATV Systems

The dB can be used to express a ratio between quantities of any similar things in a cable system; i.e., the ratio of one quantity of voltage to another, one quantity of current to another, or one quantity of power to another.

$$dB = \frac{V_1}{V_1} \qquad\qquad dB = \frac{I_1}{I_2} \qquad\qquad dB = \frac{P_1}{P_2}$$

Voltage Ratio   Current Ratio   Power Ratio

When one quantity is compared to another, V_1, I_1, or P_1 may be either smaller or greater, respectively, than V_2, I_2, or P_2. Whenever any quantity$_1$ is greater than any quantity$_2$, we precede the decibel ratio with a plus sign. Whenever any quantity$_1$ is smaller than quantity$_2$, we precede the decibel ratio with a minus sign.

When we consider the conditions normally present on a cable system, we find that quantity$_1$ will be greater than quantity$_2$ whenever there is gain and that quantity$_1$ will be less than quantity$_2$ whenever there is loss. Hence, the plus sign is always an indication of gain and the minus sign indicates loss. In practice, when it is stated that a quantity is down a quantity of dB, or that the loss is a quantity of dB, the minus sign may be omitted, but it is understood.

The important things to remember are (1) the dB is used in a cable system only to express gain or loss, and (2) the dB always expresses a ratio and never a definite quantity.

7.6 The dB as a Voltage Ratio

If a voltage ratio chart were to be made absolutely accurate, each multiplication factor would be carried out to at least five decimal places. Because measuring devices with this type of accuracy exist only in the most sophisticated laboratories and are much too expensive to be used on cable systems, our tables must be approximations of the actual values. Most test instruments used on systems are accurate within ±1 dB or more. Therefore, our table of multiplication factors is accurate enough for most purposes because the values fall within the practical limits.

There are six basic voltage ratios which are necessary to make all the dB measurements on a cable system. A complete dB-voltage conversion chart is included at the end of this chapter, but if these six ratios are memorized, there will never be a need to refer to it.

1 dB = 1.15 to 1	3 dB = 1.4 to 1	10 dB = 3.0 to 1
2 dB = 1.25 to 1	6 dB = 2.0 to 1	20 dB = 10.0 to 1

By substituting the Ohm's law, equivalent units for voltage and current, and other useful forms of the power equation are obtained.

$$P = I^2 R$$

$$P = \frac{E^2}{R}$$

Because power is usually supplied by an electric utility company or cooperative, its use represents a direct expense for the cable system. Therefore, it is important that system personnel know how to calculate power costs based on power usage rates and how to evaluate the effect of system additions or changes which would entail increased power consumption.

To calculate power cost, first total all currents supplied by each ac power supply in the system. Then, using the power equation, determine the total number of watts consumed by the system. Assume a total of 150 amps in a 60 volt system.

P = 60 volts = 150 amps

P = 9000 watts

P = 9 kilowatts

Next, determine the efficiency of the ac power supplies and adjust the total number of watts accordingly. For example, if the ac supply is 80% efficient, 80% of the power is useful energy to the system and 20% is used or consumed by the power supply itself. Therefore, the above number of watts consumed must be increased by a factor of 20%.

$$P = \frac{9\ kw}{0.8} = 11.25\ kw$$

As a final step, obtain from the power company the prevailing power usage rate and billing period interval. Assume 0.05 cents per kilowatt hour and a billing interval of 30 days. Complete the calculation based on these facts.

11.25 kw x 30 days x 24 hours per day = 8100 kwh

8100 x kwh x 0.05 cents per kwh = $405.00 each 30 days

7.7 Relationship of Voltage Ratio to Power Ratio

The fundamental concern in a CATV system is the measurement of power. The basic instrument used for measurement on a cable system is called a field strength meter and the reading obtained from it are in microvolts (μV). The instrument is actually a signal level meter and is really nothing more than a tuned volt meter. Measuring voltage on a system is not a requirement. The feature of a cable system which enables one to read power from a tuned volt meter is the fact that all system have a common impedance, which is 75 ohms Ω.

The common impedance feature means that the power in a cable system will always be directly proportional to the square of the voltage. Any power ratio can be converted to a voltage ratio by virture of this relationship.

7.8 The dB in Power Calculations

Every technician is familiar with the hybrid splitter used to split a trunk of feeder line. This device is a power splitter which does not split the voltage. However, we have shown that there is a relationship between voltage and power ratios.

The two-way splitter is designed to have 3 ½ dB insertion loss between its common terminal (the fitting where the cable to be split enters the device) and each of its other terminals (the fittings to which the two cables are connected as a result of the split).

Whatever the power entering the device, it is obvious that, if split equally, no more than one-half of this power could be present at either of the two other terminals. This means that the relationship between the power at the common fitting is to the power at either of the other fittings as two is to one, or a power ratio which can be expressed as 3 dB. If the split were a perfect one and there were no internal losses at all, this ratio would be true. However, the device cannot be made perfectly and, therefore, there is an additional one-half dB loss at the other two terminals simply because the splitter is not a perfect electronic device. Thus, this one-half dB in each leg is the insertion loss of the device.

Because the measuring device is a tuned volt meter, let us say that we have 1 volt at the entrance to the splitter. By formula, the power is proportional to the square of the voltage. With a perfect split, the power at either of the two other terminals of the device should be proportional to the square of the voltage at the common terminal divided by two. Let us say that we read 0.67 volts on our volt meter. Obviously, this is much more than 0.5, but remember that we are dealing with a power splitter in which the quantity of voltage

change is very small. Again formula, we must square this voltage and 0.67 ·
0.67 – .4489, which is certainly less than 0.5. The difference is provided by that
one-half dB of insertion loss in the device which makes our total power loss 3
½ dB instead of 3 dB.

7.9 dB as Expression of Ratio Loss

Consider a piece of RG-59/U coaxial cable, 1000′ long, a common component of most cable systems. By specification, this piece of cable will lose half of
the signal at channel 13 (with regard to voltage) by the time the signal has
passed through 100′ of cable.

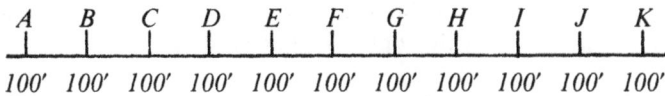

```
 A    B    C    D    E    F    G    H    I    J    K
 |    |    |    |    |    |    |    |    |    |    |
100' 100' 100' 100' 100' 100' 100' 100' 100' 100' 100'
```

Arithmetically, if we start at point A with 1 volt, we will have by specification
.5 V at point B, .25 V at point C, .125 V at point D, .0625 V at point E, .03125 V
at point F, and it complicates further as we proceed because we must divide
each time by two and take need of decimal points. Should the computations
be necessary for odd lengths of cable, it becomes even more complicated.

The calculations are simplified using the dB system. The specifications for
loss in 100′ of RG-59/U at channel 13 is given as 6 dB. Arithmetically, we had
to divide, but with dB we multiply. For two 100′ lengths of cable we multiply 6
dB by 2, so at point B we have 12 dB of loss. At point K which is 1000′ of cable,
or ten 100′ lengths, we multiply 6 dB by 10 to obtain –60 dB, which is the total
attenuation or loss in 1000 feet of RG-59/U at channel 13.

Let us suppose that we want to find the loss in 792′ of cable. This can be simply
derived using the decibel system. Because there is 6 dB of loss in 100′ of cable,
we simply divide 6 dB by 100, which gives us a loss per foot of cable of .06 dB.
Multiplying 792 · .06 dB – 47.52 dB loss at 792′.

7.10 The dB as a Gain Measurement

When we speak of gain in a cable system we are usually referring to amplifiers.
The quantity of signal at the output of an amplifier must be equal to or greater
than the quantity of signal at the input. In the case where the output is equal to
the input we have one-to-one ratio or unity gain. Unity gain simply means
that any losses contributed by the circuitry between input and output have
been compensated for by the amplifier. Whenever the output of the amplifier
is greater than the input, the ratio between the quantities is expressed in +dB
and is the gain of the amplifier. In practice, the absence of a sign prefixing dB
is understood to be plus and gain is indicated.

7.11 The dBmV

Just as 75 ohms is the common impedance for cable systems, so the accepted level which is used to compare all unknown signals is 1000 microvolts, or one millivolt measured across 75 ohms. This finite quantity is called 0 dBmV. The particular value was chosen simply because in the early days of CATV this particular level of voltage was considered a good supply voltage for a TV set.

Any signal level expressed in dBmV is either so many times greater than one millivolt or so many times less than one millivolt. Remember V_1 as used with dB. Using dBmV, V_1 is always the measured signal and V_2 is always one millivolt. Whenever V_1 is greater than V_2 the signal level is +dBmV. Whenever V_1 is less than V_2 the signal level is –dBmV. Let us use a tuneable volt meter and read the signal in microvolts at the input of a line amplifier.

$$\frac{V_1}{V_2} = \frac{3000 \ \mu V}{1000 \ \mu V} = \frac{3 \ mV}{1 \ mV}$$

This is a ratio of three-to-one, or 10 dB per millivolt, which is written 10 dBmV. This is the actual input voltage expressed in the decibel system. Let us assume that the gain of this amplifier is 20 dB. We know that the output level should be 10 times the input level, or 30,000 microvolts, or 30 mV. If the gain figure is correct, this is what we should read on the tuned volt meter ±2 mV. What is 30 mV in dBmV? Observe that 30 mV = a ratio of 30-to-one, or approximately 30 dBmV. The more simple calculation is to add 20 dB to 10 dBmV, arriving at the output of 30 dBmV. You can either add or subtract dB from dBmV and the answer will always be in dBmV, but you cannot add or subtract dBmV from dB.

Let us use a tuneable volt meter and read the signal in microvolts at the input to a typical CATV head-end control unit.

$$\frac{V_1}{V_2} = \frac{100 \ \mu V}{1000 \ \mu V} = \frac{0.1 \ mV}{1 \ mV}$$

This is a ratio of ten-to-one, but V_1 is smaller than V_2, so the ratio is expressed as –20 dBmV.

The important things to remember about dBmV are (1) the dBmV always expresses a comparison between one millivolt measured across 75 ohms (a finite quantity known as 0 dBmV), and any other smaller or greater quantity of millivolts measured across 75 ohms; (2) any quantity less than one millivolt is –dBmV; (3) any quantity greater than one millivolt is +dBmV; and (4) dB may be added to or subtracted from dBmV, but dBmV may not be added to or subtracted from dB.

Table 7.1

dBmV to +V Conversion Table
(Reference Level: 0dBmV = 1000 +V = 1mV)

dBmV	+V	dBmV	+V	dBmV	+V
-40	10.00	0	1,000	41	112,200
-39	11.22	1	1,122	42	125,900
-38	12.59	2	1,259	43	141,300
-37	14.13	3	1,413	44	158,500
-36	15.85	4	1,585	45	177,800
-35	17.78	5	1,778	46	199,500
-34	19.95	6	1,995	47	223,900
-33	22.39	7	2,239	48	251,200
-32	25.12	8	2,512	49	281,800
-31	28.18	9	2,818	50	316,200
-30	31.62	10	3,162	51	354,800
-29	35.48	11	3,548	52	398,100
-28	39.81	12	3,981	53	446,700
-27	44.67	13	4,467	54	501,200
-26	50.12	14	5,012	55	562,300
-25	56.23	15	5,623	56	631,000
-24	63.10	16	6,310	57	707,900
-23	70.79	17	7,079	58	794,300
-22	79.43	18	7,943	59	891,300
-21	89.13	19	8,913	60	1,000,000
-20	100.00	20	10,000	61	1,122,000
-19	112.2	21	11,220	62	1,259,000
-18	125.9	22	12,590	63	1,413,000
-17	141.3	23	14,130	64	1,585,000
-16	158.5	24	15,850	65	1,778,000
-15	177.8	25	17,780	66	1,995,000
-14	199.5	26	19,950	67	2,239,000
-13	223.9	27	22,390	68	2,512,000
-12	251.2	28	25,120	69	2,818,000
-11	281.8	29	28,180	70	3,162,000
-10	316.2	30	31,620	71	3,548,000
- 9	354.8	31	35,480	72	3,981,000
- 8	398.1	32	39,810	73	4,467,000
- 7	446.7	33	44,670	74	5,012,000
- 6	501.2	34	50,120	75	5,623,000
- 5	562.3	35	56,230	76	6,310,000
- 4	631.0	36	63,100	77	7,079,000
- 3	707.9	37	70,790	78	7,943,000
- 2	794.3	38	79,430	79	8,913,000
- 1	891.3	39	89,130	80	10,000,000
0	1,000.0	40	100,000		

Table 7.2

dBmV to dBm Conversion Table

(Reference Level: 0dBmV = –48.75dBm)

(0dBm = 1 mW)

–40	–88.75	0	–48.75	41	– 7.75
–39	–87.75	1	–47.75	42	– 6.75
–38	–86.75	2	–46.75	43	– 5.75
–37	–85.75	3	–45.75	44	– 4.75
–36	–84.75	4	–44.75	45	– 3.75
–35	–83.75	5	–43.75	46	– 2.75
–34	–82.75	6	–42.75	47	– 1.75
–33	–81.75	7	–41.75	48	– 0.75
–32	–80.75	8	–40.75	49	0.25
–31	–79.75	9	–39.75	50	1.25
–30	–78.75	10	–38.75	51	2.25
–29	–77.75	11	–37.75	52	3.25
–28	–76.75	12	–36.75	53	4.25
–27	–75.75	13	–35.75	54	5.25
–26	–74.75	14	–34.75	55	6.25
–25	–73.75	15	–33.75	56	7.25
–24	–72.75	16	–32.75	57	8.25
–23	–71.75	17	–31.75	58	9.25
–22	–70.75	18	–30.75	59	10.25
–21	–69.75	19	–29.75	60	11.25
–20	–68.75	20	–28.75	61	12.25
–19	–67.75	21	–27.75	62	13.25
–18	–66.75	22	–26.75	63	14.25
–17	–65.75	23	–25.75	64	15.25
–16	–64.75	24	–24.75	65	16.25
–15	–63.75	25	–23.75	66	17.25
–14	–62.75	26	–22.75	67	18.25
–13	–61.75	27	–21.75	68	19.25
–12	–60.75	28	–20.75	69	20.25
–11	–59.75	29	–19.75	70	21.25
–10	–58.75	30	–18.75	71	22.25
– 9	–57.75	31	–17.75	72	23.25
– 8	–56.75	32	–16.75	73	24.25
– 7	–55.75	33	–15.75	74	25.25
– 6	–54.75	34	–14.75	75	26.25
– 5	–53.75	35	–13.75	76	27.25
– 4	–52.75	36	–12.75	77	28.25
– 3	–51.75	37	–11.75	78	29.25
– 2	–50.75	38	–10.75	79	30.25
– 1	–49.75	39	– 9.75	80	31.25
0	–48.75	40	– 8.75		

NUMERIC VALUE 1 TO 10: LOG 0 TO 1: $10^0 = 1$, $10^1 = 10$

NUMERIC VALUE 10 TO 100: LOG 1 TO 2: $10^2 = 100$

NUMERIC VALUE 100 TO 1000: LOG 2 TO 3: $10^3 = 1000$

EXAMPLES:
  LOG OF   2 = 0.301
  LOG OF 20 = 1.301
  LOG OF 200 = 2.301

POWER RATIO IN dB = 10 LOG $\frac{P_1}{P_2}$

•VOLTAGE RATIO IN dB = 20 LOG $\frac{E_1}{E_2}$

•WHERE E_1 AND E_2 ARE ACROSS SAME IMPEDANCE.

0 dBmV = 1 mV ACROSS 75 OHMS

0 dBm =   1 mW ACROSS 50 OHMS
             224 mV ACROSS 50 OHMS

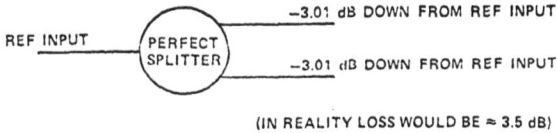

Figure 7.2 Logarithms

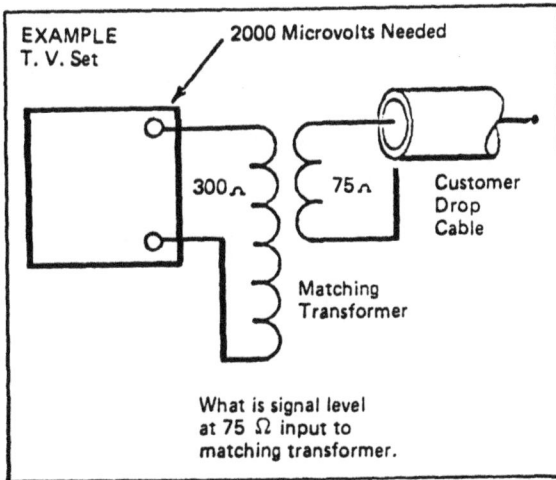

Figure 7.3 Example TV set

8 System Performance

The following is a discussion of recommended performance specifications used in system designs along with tables that give the recommended levels to achieve this performance.

8.1 Carrier-to-Noise Ratio (C/N)

The carrier-to-noise ratio is defined as the ratio between the video carrier signal level and the RMS noise level. The ratio is expressed in decibels. The threshold of perceptibility of noise on a television receiver occurs at a C/N ratio of approximately 47 dB. Reasonable picture quality can be achieved with a C/N ratio of about 43 dB on some television sets.

8.2 Carrier-to-Cross Modulation Ratio (C/X)

Cross modulation is defined as the third order distortion which causes the modulation from one signal carrier to modulate another signal carrier. Cross modulation may be defined in other ways, but the CATV definition is dictated by the NCTA method of measurement. The carrier-to-cross modulation ratio is measured by using 15.75 kHz synchronous modulation, and a 50% duty cycle. This is a worst case measurement which does not usually occur in a system since the sync is random on the various channels transmitted in a system. The threshold of perceptibility of this ratio on a television receiver is less than 40 dB. The C/X ratio is not the limiting factor in the design of most systems and is, therefore, below the threshold in system designs.

8.3 Carrier-to-Third Order Intermodulation Ratio (C/IM)

Third-order intermodulation is the simultaneous pulse of two or three signal carriers to produce a spurious carrier, caused by the third-order distortion characteristics of the amplifier. The simultaneous pulse, or beating together, of three carriers is commonly called "triple beat."

Figure 8.1 shows the threshold of perceptibility of an individual triple beat (curve 1), or the summation of the total triple beats on a channel (curve 2, composite triple beat) *versus* the number of triple beats per channel. There is some margin in the curve because modulation on the carriers reduces the average level of a video carrier by approximately 5 dB, which represents a 10 dB improvement in the carrier-to-intermodulation ratio.

Composite triple beat is the summation of spurious signals generated by the third-order transfer characteristic of active devices (i.e., transistors). These spurious signals "stack" together on each channel, causing interference in the television picture if the level is too high.

In a 30-channel system, approximately 229 spurious signals (triple beats) stack on the center channel (i.e., channel 8). These beats are functions of all 30 channels and vary randomly relative to each other. This type of third-order distortion is generally the limiting factor in the output capability of an amplifier.

Correlation of specifications between manufacturers has become a problem

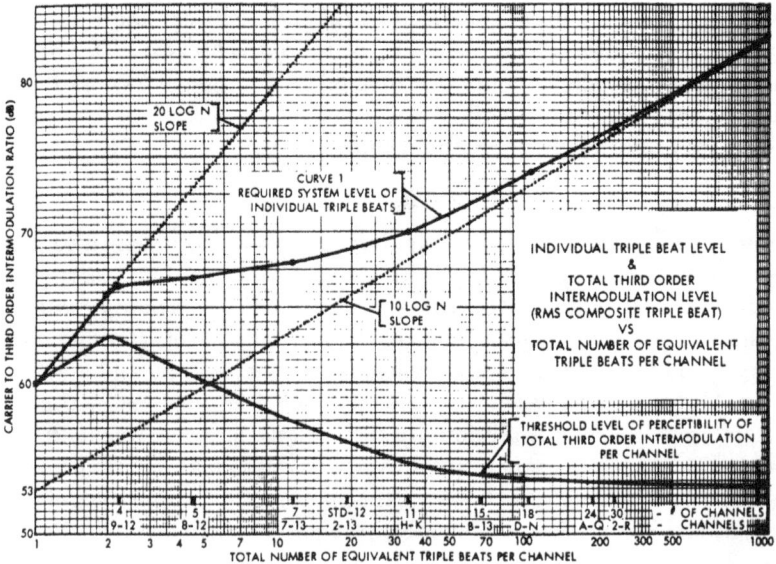

Figure 8.1 Individual Triple Beat Level and Total Third-Order Intermodulation Level vs. Total Number of Equivalent Triple Beats per Channel
Source: RCA Cable Construction Manual

due to different methods to measure composite triple beat. Actually the specifications are the same when the method of testing is the same. The following is a list of the parameters that one must be aware of when specifying composite triple beat:

Number of Channels — The number of triple beats per channel increases exponentially as the number of channels increases. A difference of 2 dB in the specification will be observed when the number of channels is increased from 30 to 35.

Levels — Because triple beat is a third-order distortion, the distortion will increase in output level. If the amplifier operates with a tilt, the distortion will also be affected. A 6 dB output tilt will give approximately 4 dB improvement in the carrier-to-distortion ratio over a flat output.

Measurement — Normally, a random noise (which the composite triple beat approximates), will be specified as an RMS value. Other manufacturers have used instruments other than a true RMS volt meter to measure the triple beat and have not corrected the reading back to an RMS reading. For example, the spectrum analyzer in the "log" mode will read approximately 3 dB better than true RMS. The Jerrold SLM will read approximately 9 dB worse than true RMS for a 30-channel system.

Modulated or Unmodulated Carriers — A good engineering practice is to specify the measurement of triple beat with unmodulated carriers, because the distortion level will depend upon the type of modulations used. RCA, a manufacturer of cable TV equipment, published a paper entitled "Composite Triple Beat Measurements." In the RCA paper, a calculation assuming a modulation with a uniform distribution of grey-level information is made, showing that the average level of a carrier is reduced approximately 5 dB from the unmodulated case, thereby leading to a distortion reduction of 10 dB. Experiments were made to verify this calculation. Some other manufacturers have chosen a 3 dB factor for modulated video carrier level when measuring triple beat and have added this number to their specification, hence reducing their distortion specification by 6 dB.

8.4 Carrier to Second-Order Intermodulation Ratio

Carrier to second-order intermodulation is the simultaneous pulse, or beating together, of two signal carriers because of the second-order distortion characteristics of the amplifier. Second-order intermodulation would follow the same curves in Fig. 8.1 for perceptibility, except that there are considerably fewer spurious signals which would sum together on the one channel. All second-order beats fall ± 0.75 and ± 1.25 MHz about the video carrier. The beats below the carrier are cut off by the filtering in the TV receiver and, therefore, are not significant.

8.5 System Performance Relationships

To evaluate the cascade performance of identical components (i.e., truck amplifiers, bridging amplifiers, or line extender amplifiers) use the following relationships:

$$C/N \text{ (single amp)} = Lo - (Nt + Nf + G) \tag{8-1}$$

where

C/N = Carrier-to-noise ratio

Lo = Amplifier operational output

Nt = Thermal noise (-59 dBmV in a 75 ohm system at a temperature of 68°F and measurement bandwidth of 4 MHz)

NF = Amplifier noise figure

G = Amplifier operational gain

and all quantities are expressed in decibels (dB).

For second-order distortion:

C/N (cascade) = C/N (single amp) − $10 \log_{10} N$

NF (cascade) = NF (single amp) + $10 \log_{10} N$

X-Mod (cascade) = *X-Mod* (single amp) + $20 \log_{10} N$

2nd Order (cascade) = *2nd Order* (single amp) + $10 \log_{10} N$

Single Triple Beat (cascade) = *STB* (single amp) + $20 \log_{10} N$

Composite Triple Beat (cascade) = *CTB* (single amp) + $20 \log_{10} N$

Hum Mod (cascade) = *Hum Mod* (single amp) + $20 \log_{10} N$

where N = number of identical amps in cascade, and all quantities are expressed in decibels.

The expression for second-order is an approximate based on long cascades. Actual measured second-order can be greater or less than the calculated value. The amount of deviation from the calculated value will statistically increase as the cascade length decreases.

The formula given above for second-order distortion is theoretical. In actual system operation, the more practical form might be as follows:

2nd Order (cascade) − *2nd Order* (single amp) + $15 \log_{10} N$

The above expressions permit cascade calculations for amplifiers which operate at the output level specified by the manufacturer. When it is necessary to operate an amplifier at higher or lower levels, these expressions must be modified. The term $2(Lo - Lr)$ is inserted in the equations for cross modulation (X-Mod) and triple beat (TB) provides the necessary equation modifications for calculations at levels other than those for which the amplifiers are rated.

$$2\,(Lo - Lr)$$

$$X\text{-}Mod \text{ (cascade)} = X\text{-}Mod \text{ (single amp)} + 2(Lo - Lr) + 20 \log_{10} N$$

$$TB \text{ (cascade)} = TB \text{ (single amp)} + 2\,(Lo - Lr) + 20 \log_{10} N$$

The following term $(Lo - Lr)$ inserted in the equation for C/N and second-order provides the necessary equation modifications for calculations at levels other than those for which the amplifiers are rated. Thusly, we have

$$(Lo - Lr)$$

$$2nd\ Order \text{ (cascade)} = 2nd\ Order \text{ (single amp)} + (Lo - Lr) + 20 \log_{10} N$$

$$C/N \text{ (cascade)} = C/N \text{ (single amp)} + (Lo - Lr) - 10 \log_{10} N$$

where

$$Lo = \text{Actual amplifier output level}$$

$$Lr = \text{Rated amplifier output level}$$

To evaluate overall system performance, convert the appropriate cascade performance numbers [Eq. (8-1)] to equivalent power or voltage ratios and calculate the sum. Then convert the sum of the voltage ratios to corresponding decibel values.

For overall system second-order and C/N calculations use:

$$\text{dB (system)} = -10 \log_{10} \left(\frac{-\text{dB}\,(TK)}{10} \quad \frac{-\text{dB}\,(Br)}{10} \quad \frac{-\text{dB}\,(le)}{10} \right) \qquad (8\text{-}2)$$

For overall system X-Mod, TB, and $Hum\ Mod$ calculations use:

$$\text{dB (system)} = 20 \log_{10} \left(\frac{-\text{dB}\,(TK)}{20} \quad \frac{-\text{dB}\,(Br)}{20} \quad \frac{-\text{dB}\,(le)}{20} \right) \qquad (8\text{-}3)$$

where

dB (*TK*) = Cascade distortion factor for trunk

dB (*Br*)   = Cascade distortion factor for bridger

dB (*Le*)   = Cascade distortion factor for line extender

$$(8-3)$$

To determine the required amplifier input level (identical amplifiers) for a given cascade, C/N use the following expression:

$$Lin = C/N + Nt + NF + 10 \log_{10} N \qquad (8-4)$$

where

C/N = Desired cascade C/N

Nt = Thermal noise (–59 dBmV at 68°F, 75 ohms, 4 MHz bandwidth)

NF = Amplifier noise figure

N = Number of identical amps in cascade

To determine the required amplifier output level (identical amplifiers) for a given cascade C/N and *X-Mod*, use the following expression:

$$Lo = Lr + \frac{(X\text{-}Mod)\,c - (X\text{-}Mod)r - 20 \log N}{4} \qquad (8-5)$$

$$+ \frac{(C/N)\,c + Nt + F + G - Lr + 10 \log N}{2}$$

where

(*X-Mod*) *r* = Amplifier rated *X-Mod*

(*X-Mod*) *c* = Desired cascade *X-Mod*

(*C/N*) *c* = Desired cascade *C/N*

Nt = Thermal noise (–59 dBmV at 68°F, 75 ohms, 4 MHz bandwidth)

NF = Amplifier noise figure

G = Amplifier operational gain

Lr = Amplifier rated output level

N = Number of identical amps in cascade

9 Measurement Parameters

Service and repair of faulty equipment is one important aspect of the operator's distribution system responsibilities. The other is preventive maintenance, the so-called "proof of performance" testing which qualifies a system for regulatory (government) agency approval.

For a complete understanding of proof of performance measurements, an overview of electronic circuit measurement techniques follows. Whether the electronics is a circuit board in a trunk amplifier or an entire distribution system, it can be thought of as a "black box" with one or more connectors for access.

Two approaches can be taken to determine how this mysterious black box functions. Either monitor each of the connectors while the box performs in its normal capacity, or feed a known signal in one connector and compare this signal very accurately with the output at another connector. The first procedure is called *signal analysis*. The second, because it tells specifically how much the black box circuit changes a signal, is called *network analysis* (see Fig. 9.1).

Signal analysis measurements are relatively simple. Measuring the dc voltages of a bridger's power supply, the frequency of a converted UHF channel at the head-end, and the system noise level at a subscriber drop are examples of signal analysis. Instruments which measure amplitude, frequency, and amplitude-frequency combinations are required for signal analysis.

Network analysis measurements offer more information about a circuit or system, but the techniques and equipment are more complicated than a signal analysis. Group delay is an example of network analysis measurement. Equipment which can provide a test signal as well as measure frequency, amplitude, phase, and their combinations are usually required for network analysis.

(a) NETWORK ANALYSIS

(b) SIGNAL ANALYSIS

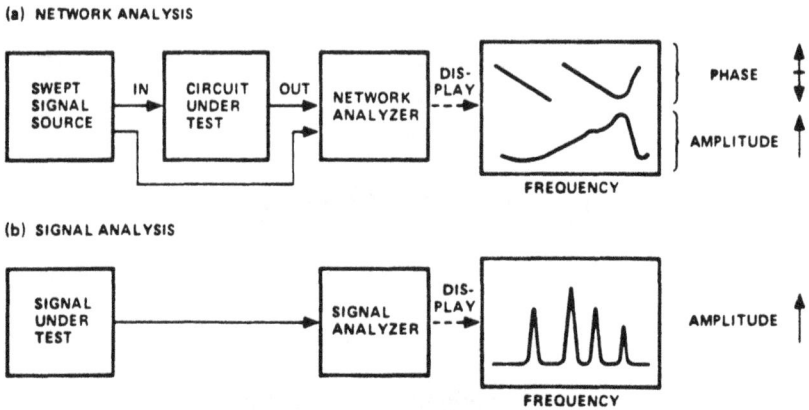

Figure 9.1 Network Analysis and Signal Analysis Note: The difference between these techniques is what they measure: network analysis is characterizing a circuit by displaying the phase shift and amplitude changes on a known input spectrum (a swept frequency source); whereas signal analysis displays the amplitude spectrum of a test signal. The displays in both methods may be any one of a number of readouts: CRT, taut-band meter, digital meter and others.

Source: Hewlett Packard, Cable Television System Measurement Handbook

Clearly the common denominator in these measurement techniques are the parameters of frequency, amplitude, phase, and various combinations. Let us summarize how these parameters are best qualified for the CATV industry.

9.1 Signal Frequency

The aspect of frequency which is most important to CATV systems is *accuracy*. Inaccurate carrier frequencies or signal spacing can cause serious distortions and interference as vital sidebands are absorbed by passive bandpass filters or infringe upon signals from neighboring channels.

Signal accuracy is almost always expressed as a percent of the radio frequency (RF) parts per million, or in hertz. In each case, it is understood that the accuracy figure gives the limits of the tolerance range. For example, a television carrier at 193.25 MHz must be within 100 parts in 10^6 for a broadcast power of (LTQ) 100 watts or 1000 Hz for a power of $>$ 100 watts.

$$100 \text{ watts} = 193.25 \text{ MHz} \pm (193.25 \text{ MHz} \times 100 \times 10^{-6}), \text{ or}$$

$$193.25 \text{ MHz} \pm 19.325 \text{ kHz or } 193.25 \text{ MHz} \pm 0.01\%$$

$$100 \text{ watts} = 193.25 \text{ MHz} \pm 1000 \text{ Hz, or}$$

$$193.25 \text{ MHz} \pm 0.00052\% \tag{9-1}$$

Frequency stability is another important consideration. Frequency stability refers to the tendency of signal sources to drift with time, temperature, electrical interference, and mechanical vibration. Short-term frequency deviation is called residual frequency modulation (FM), and long-term change is called drift. An example of residual FM would be 100 Hz peak-to-peak. This means that with any tendency of the carrier to swing rapidly above and below its normal frequency, it will not swing more than 100 Hz either up or down in any one swing. Drift may be 100 parts per million in 12 hours, which means for a carrier, f_c, the frequency change f_c, the frequency change f_c will be less than $\pm 100/10^6 \cdot f_c$ Hz (where f_c is in hertz) in 12 hours.

For the measurement of frequencies whose accuracy and stability are suspect, the measurement instrument should have better frequency accuracy and stability than specified by at least a factor of three. In any CATV system, such an instrument could pinpoint carrier and pilot inaccuracies (see Fig. 9.2), assure correct channel and sideband spacing, and identify unwanted signals that may be interfering with the system.

9.2 Signal Level

Throughout the CATV system, power is distributed in the form of TV and FM carriers, pilot tones, test signals, dc power supply, and noise. Specified levels must be maintained at each point in the system to ensure good performance, and the levels at different frequencies are also significant. Measurement techniques are given below.

Let us start with the cable. The characteristic impedance of distribution cable is 75 ohms. This impedance is the amount of resistance that the cable signals can "see" from the center conductor to the other shield of the cable. All of the signal voltage and currents traveling along the cable are governed by this impedance simply by virture of Ohm's law. The rms voltage open the cable is related to the power transmitted by

$$P = \frac{V^2}{R} = \frac{V^2}{75}$$

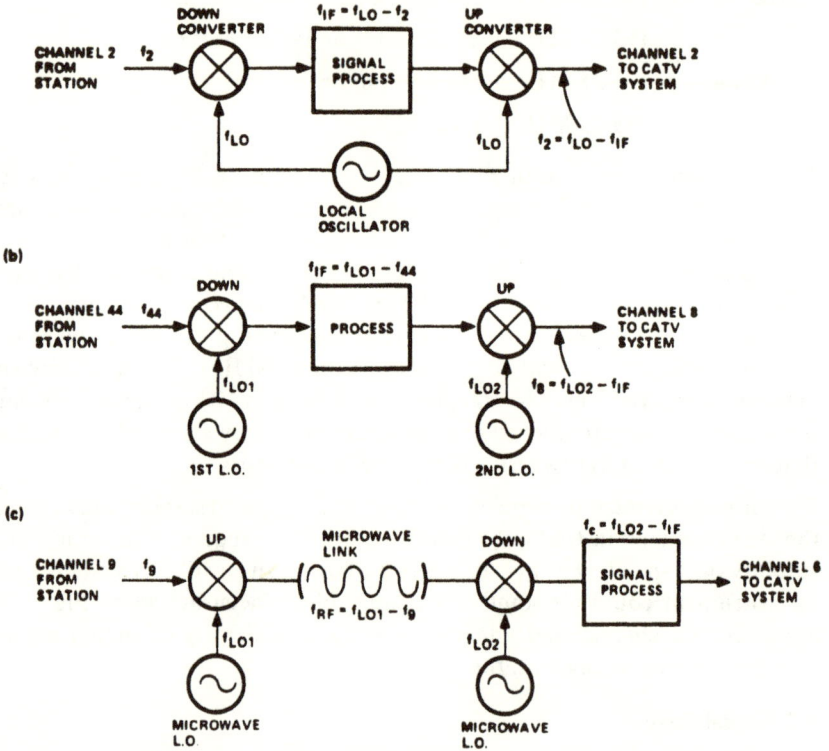

Figure 9.2 Frequency Accuracy and Stability. Note: These three typical conversion schemes illustrate where errors could be introduced: (a) IF signal processing; (b) When a channel is converted to another slot the two conversions are independent and deviation in either LO will show up as a change in output frequency; (c) Simplified microwave relay conversion process where the two LOs are independent.

Source: Hewlett Packard, Cable Television System Measurement Handbook

where

P = signal power in watts

V = signal voltage in volts rms

R = cable impedance in ohms   (9-2)

Many CATV system measurements involve differences in signal power. Voltage differential is an inefficient measure of the power differential because each time a power change is measured, the formula $(V_1^2 - V_2^2)/75\Omega$ must be calculated.

The decibel form of measurement (i.e., ratio) resolves these difficulties in handling system power figures. It is defined as

$$dB = 10 \log_{10} \frac{P_2}{P_1}$$

where dB = decibel.

$\dfrac{P_2}{P_1}$ = ratio of two powers, and P_1 usually the reference. Since

$$P_1 = \frac{V_1^2}{75} \quad \text{and} \quad P_2 = \frac{V_2^2}{75}$$

the dB can be expressed as a ratio of voltages

$$dB = 10 \log \frac{V_2^2/75}{V_1^2/75} = 10 \log \frac{V_2^2}{V_1^2} = 20 \log \frac{V_2^2}{V_1} \qquad (9\text{-}3)$$

where

$\dfrac{V_2}{V_1}$ = ratio of two voltages.

$$(9\text{-}3)$$

If V_1 is defined as the reference and set equal to 1 mV (10^{-3} volts), then the dB can be called a "decibel referred to one millivolt," or dBmV.

$$dBm\,V = 20 \log \frac{V}{10^{-3}} \qquad (9\text{-}4)$$

Note that the characteristic impedance is not indicated in the formula, and it need not be, as long as the computations are for powers and voltages in systems of the same impedance. Here are some examples of converting voltage to dBmV:

$$1\,V = 20 \log \frac{1}{10^{-3}} = 60 \text{ dBm V}$$

$$0.1\,V = 20 \log \frac{10^{-1}}{10^{-3}} = 40 \text{ dBm V}$$

Table 9.1
dBmV Impedance Conversion

FROM dBmV IN Z_1	TO dBmV IN Z_2				
	50Ω	75Ω	300Ω	600Ω	Z_2
50Ω	0	+1.76 dB	+7.78 dB	+10.79 dB	$+10 \log \dfrac{Z_2}{50}$
75Ω	–1.76 dB	0	+6.02 dB	+9.03 dB	$+10 \log \dfrac{Z_2}{75}$
300Ω	–7.78 dB	–6.02 dB	0	+3.01 dB	$+10 \log \dfrac{Z_2}{300}$
600Ω	–10.79 dB	–9.03 dB	–3.01	0	$+10 \log \dfrac{Z_2}{600}$
Z_1	$+10 \log_{10} \dfrac{50}{Z_1}$	$+10 \log_{10} \dfrac{75}{Z_1}$	$+10 \log_{10} \dfrac{300}{Z_1}$	$+10 \log_{10} \dfrac{600}{Z_1}$	$+10 \log_{10} \dfrac{Z_2}{Z_1}$

Note: If the power in a systrem is uniform, these conversion factors will adjust the dBmV reading for different cable and circuit impedances. Note that the insertion loss of impedance matching transformer is not included.

$$0.01 \; V = 20 \log \frac{10^{-2}}{10^{-3}} = 20 \, dBm \; V$$

$$10 \; \mu V = 20 \log \frac{10^{-5}}{10^{-3}} = -40 \, dBm \; V$$

Two other units commonly used to measure RF power are dBm and dBV, dB above a milliwatt (10^{-3} watts) and dB above a microvolt (10^{-6} volts), respectively. Table 9.1 gives dBmV impedance and Table 9.2 describes the conversion from each unit to the other.

The important attribute common to each of these units is their relationship to power. A change in dB, no matter what unit referred to (volts, microvolts, or milliwatts), denotes the same power change. A 3 dB increase (or decrease) in a signal level means its power has doubled (or halved). From the dB power definition, we have

$$dB = 10 \log \frac{P_2}{P_1} = 20 \log_{10} \frac{V_2}{V_1} \tag{9-5}$$

$$3 \, dB = 10 \log 2$$

To understand the relationship of dB to power ratio, a table of the more common power ratios would be assembled as follows:

To derive other power ratios from Table 9.3, simply add the combination of dB from the left column and multiply the corresponding ratios in the right column.

Now let us consider the various forms of signal level measurement in CATV systems and their importance. In Fig. 9.3, typical dBmV levels are shown at various points throughout a distribution line. These levels represent the TV video carrier peak signal. The peak of the video carrier is simply the un-modulated carrier voltage, as shown in Fig. 9.3. The peak detection value in Fig. 9.3(c) is the only meaningful signal level measurement for this type of signal because none of the envelope information is lost. Each visual carrier at the subscriber's drop should be at a given power level and within a given number of dB from each other for proper receiver performance. When considering the TV channels as a spectral display, that is, a display of signal amplitude *versus* frequency, these specifications are more clearly understood. Figure 9.4 shows a simplified CATV spectrum where the nominally absolute signal level is 0 dBmV. However, the end points, channel 2 and channel 6 are six dBmV, which is a power differential factor of four. The home TV receiver displays the weaker signal, with "snow" and poor color quality, whereas the stronger channel 2 would be considerably less noisy.

Conversion Formula for Common Units of Signal Level Measurement

	TO				
	VOLTS, V	**WATTS, W**	**dBm**	**dBmV**	**dBµV**
VOLTS, V	V	$\dfrac{V^2}{Z}$	$10\log\dfrac{V^2}{10^{-3}\times Z}$	$20\log\dfrac{V}{10^{-3}}$	$20\log\dfrac{V}{10^{-6}}$
WATTS, P	$\sqrt{P\times Z}$	P	$10\log\dfrac{P}{10^{-3}}$	$20\log\left(\dfrac{\sqrt{Z\times P}}{10^{-3}}\right)$	$20\log\left(\dfrac{\sqrt{Z\times P}}{10^{-6}}\right)$
dBm	$\left(\log^{-1}\left[\dfrac{dBm}{10}\right]\times Z\times10^{-3}\right)^{1/2}$	$\log^{-1}\left[\dfrac{dBm}{10}\right]\times10^{-3}$	dBm	$dBm + 30$ $+20\log\sqrt{Z}$	$dBm + 90$ $+20\log\sqrt{Z}$
dBmV	$\log^{-1}\left[\dfrac{dBmV}{20}\right]\times10^{-3}$	$\left(\log^{-1}\left[\dfrac{dBmV}{20}\right]\right)^2\times\dfrac{10^{-6}}{Z}$	$dBmV - 30$ $-20\log\sqrt{Z}$	dBmV	$dB\mu V + 60$
dBµV	$\log^{-1}\left[\dfrac{dB\mu V}{20}\right]\times10^{-6}$	$\left(\log^{-1}\left[\dfrac{dB\mu V}{20}\right]\right)^2\times\dfrac{10^{-12}}{Z}$	$dB\mu V - 90$ $-20\log\sqrt{Z}$	$dB\mu V - 60$	dBµV

Z = the impedance of the system

Common conversion factors:

1 milliwatt (10^{-3} watts) in 75Ω = 0 dBm = 48.75 dBmV

0 dBm = +107 dBµV in 50Ω

0 dBmV = +60 dBµV

0 dBmV (50Ω) = +1.75 dBmV (75Ω) See Table 2.1

Note: Find the units of the value to translate in the left column, then use the corresponding formula under the units column desired. Example: Convert +10 dBm to dBmV in 75Ω. Substitute into the formula "dBm + 30 + 20 log $\sqrt{z}$", 10 dBm = +10 +30 +20 log $\sqrt{75}$ = 40 + 18.75 = 58.75 dBmV (75Ω). If we now wished to convert this figure into 50Ω we would refer to the left column of Table 2.1 to find the 75Ω row, and find the conversion factor under the desired 50Ω, −1.76 dB to give 58.75 dBmV (75Ω) = 58.75 −1.75 = 57.0 dBmV (50Ω). Note that mismatch and insertion losses are not included in these conversions.

Table 9.3

dB (Add)	$\dfrac{P_2}{P_1}$ (Multiply)	dB (Add)	$\dfrac{P_2}{P_1}$ (Multiply)
0	1	−1	0.8
1	1.25	−3	0.5
3	2	−6	0.25
5	3.16 (= $\sqrt{10}$)	−9	0.125
6	4	−10	10^{-1}
9	8	−20	10^{-2}
10	10	−30	10^{-3}
20	100		
30	1000		

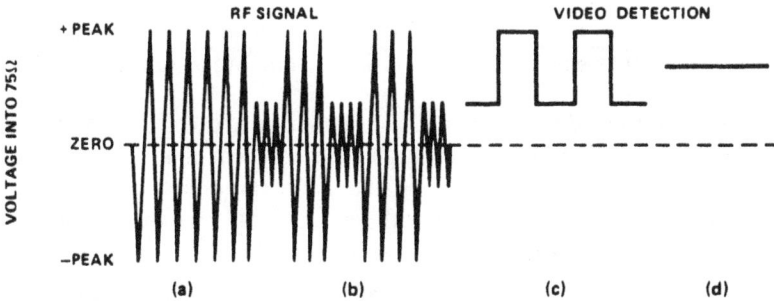

Figure 9.3 Simplified TV Carrier Signal: (a) unmodulated, (b) modulated, (c) detected (stripped of RF signal) for peak envelope response, and (d) detected for average envelope response. The maximum RF signal level (peak) is the signal voltage proportional to the power of the signal since the impedance of the system is uniform 75 ohms.

Source: Hewlett Packard, Cable Television System Measurement Handbook

This power level *versus* frequency parameter is called *flatness*. A flat system is one which will perfectly reproduce the power level *versus* frequency profile of a swept signal put in at the head-end. It needs to be specified along with absolute signal level to prevent wide variations in picture quality: too small a signal would cause the picture to drop out; and too strong a signal would

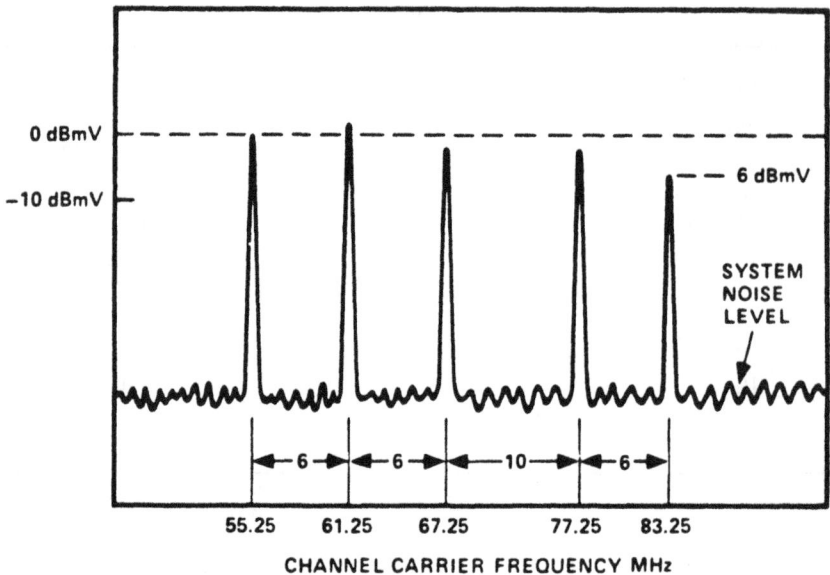

Figure 9.4 Spectral Display of TV Carriers for Channels 2 through 6.
Source: Hewlett Packard, Cable Television System Measurement Handbook

cause compression, adjacent channel, and radiation problems. Complementing the flatness specification are the audio and adjacent channel level specifications, which insure that the various sideband components of the complex channel spectrum will not interfere with one another. Note that a system can be flat and still violate the adjacent channel specifications.

In Figure 9.5, channel 2 and channel 3 are presented along with their audio sidebands (with no modulation complicating the figure). The adjacent signal level must be within ±3 dB and the audio carrier must be 13 to 17 dB lower. This will insure that the channel 3 video will be at least 10 dB above the channel 2 audio, preventing audio modulation occuring on the channel 3 picture.

Control of the CATV system's flatness is accomplished with automatic gain control (AGC) and automatic slope control (ASC) through the use of pilot tones inserted at two or more strategic frequencies. Cable losses are higher at higher frequencies, so the flatness or frequency response of a cable looks like the illustration in Fig. 9.6. At 300 MHz, a 1500′ cable can lose as much as 18 dB more than would occur in transmission at 54 MHz. To compensate for this greater loss, the amplifiers along the system shape their gain response with adequate tilt or slope to boost the high frequency end of the signal. Pilot signal levels, and sometimes pilot sidebands, act as standards by which the

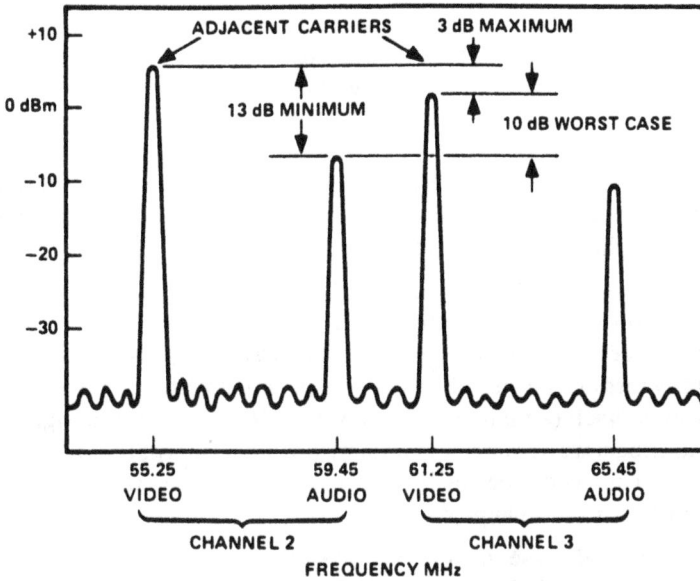

Figure 9.5 Adjacent Carriers with their Audio Sidebands. Note: Signal level specifications insure minimal interference between the audio of one channel and the video of the adjacent channel.

Source: Hewlett Packard, Cable Television System Measurement Handbook

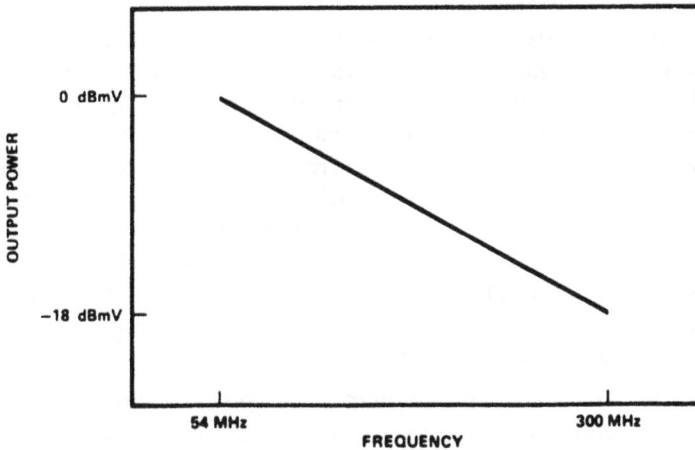

Figure 9.6 Typical CATV Trunk Cable Frequency Response for 1500 ft. of Cable

Source: Hewlett Packard, Cable Television System Measurement Handbook

system monitors itself, then adjusts AGC and ASC levels accordingly. Specifications for these pilots are set by the various equipment manufacturers and require checking as part of routine system maintenance.

9.3 Noise

The smooth, orderly flow of the electrons of an RF signal can be disrupted by another type of signal energy whose electron flow in random. This randomness, caused by heat's action on resistive elements, is *noise.*

Noise has all the attributes of a signal; it has a level and frequency response and it can be amplified, transmitted, and measured. Noise is detrimental to a CATV system because it distorts or obliterates desired signals. Noise does not only amplify through an amplifier, but is disproportionately increased by the amplifier noise itself (amplifier noise power, or NP). In system design, noise figure is an expensive parameter to minimize, but the alternative of maximizing signal level power can be much more costly. Noise figure will be covered in more detail later in this chapter.

The randomness of noise gives it theoretically infinite frequency response. Figure 9.7 shows an RF signal embedded in noise in both the time and frequency domains. At any one point along either the time or frequency axis, noise appears as a range of amplitudes rather than a single value. The attributes of wide frequency response and amplitude range are the basis for noise measurement. If the amplitude range is averaged over a specified frequency range in units of power per Hz, a meaningful measurement results, called the *noise power density.*

CATV noise measurements are always referred to a standard frequency window, called a bandwidth, so the readings are consistent. The wider the bandwidth used, the more noise power density detected by the measuring instrument and the higher the noise signal power. Very narrow bandwidths pass less of this random noise signal. Therefore, narrow bandwidths are used for measurements of small RF signals, which could easily be buried in noise. However, CATV measurements of noise must be referred to a bandwidth representative of the equipment in the system. The 6 MHz TV channel is the common denominator. Of this 6 MHz, most of the picture information is contained in a 4 MHz bandwidth. Hence, the noise measurements for CATV are referred to a 4 MHz bandwidth.

For noise level measurements not made with a 4 MHz bandwidth, a simple mathematical conversion is necessary. The noise power (in dBm or dBmV) changes as the ratio of the bandwidth:

$$NP = 10 \log_{10} \frac{BW_1}{BW_2} \tag{9-6}$$

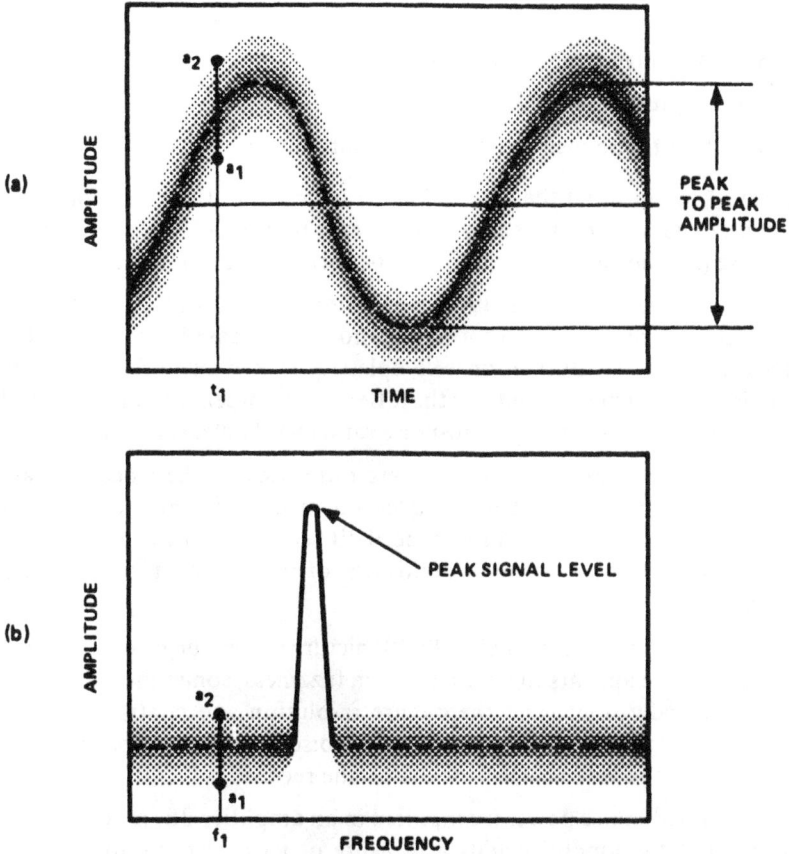

Figure 9.7 Signal and Noise in the Time and Frequency Domains: (a) a sinusoidal signal in the time domain whose waveform is thickened by the random amplitude variations of noise. The dashed line represents the average of the noise signal, the sine wave itself. (b) In the frequency domain the same RF signal shows as a spike whose amplitude is representative of the signal level.

Source: Hewlett Packard, Cable Television System Measurement Handbook

where

NP = change in noise power in dB

BW_1 = reference bandwidth in Hz

BW_2 = test data measurement bandwidth in Hz

Always keep in mind that the wider the bandwidth, the more noise power which can pass. For our example, this means that the –25 dBmV figure must increase (i.e., become less negative) to be corrected for the wider bandwidth.

Noise is the hazard that can obscure or distort a signal, as Fig. 9.8 illustrates. In the graphs from top (a) to bottom (b), noise is increased until the amplitude modulation (AM) information is completely masked by noise. Another type of noise measurement quantifies the difference between the noise and the RF signal peak, that is, the signal-to-noise ratio (SNR) measurement.

In CATV, this is called carrier-to-noise ratio because the video TV carrier level compared to the system noise level is measured. This carrier-to-noise ratio (C/N) is measured in unreferenced dB because it is a ratio of two levels. Figure 9.8(a) shows an example of this type of measurement in the frequency domain.

When the C/N ratio approaches 40 dB, picture quality begins to degrade. A "noisy" picture appears to have a random fuzziness, sometimes called *snow*, that graphically obliterates the picture resolution and contrast. As in the simple AM example of Fig. 9.8, as the noise level increases, the various sidebands of the video carrier are lost to the receiver.

As mentioned, noise power is amplified in an amplifier dB, for dB with CW signals, but the amplifier adds a measure of its own noise to the output because of the inherent noise of active devices such as transistors. The measure of the amplifier characteristic is noise figure, or NF. Table 9.4 and Fig. 9.9 show the relationship of noise power density, amplifier gain, and noise figure.

9.4 Interference

Any signal present within the passband or the TV channel which causes a degradation of the receiver's quality is called an interference signal. This interference signal's source may be originated outside the system (co-channel), generated within the system (intermodulation, hum, and cross modulation), or produced by the system (radiation). Let us take these one category at a time, looking first at the signal principles of each, then relating interference signal levels to reception quality.

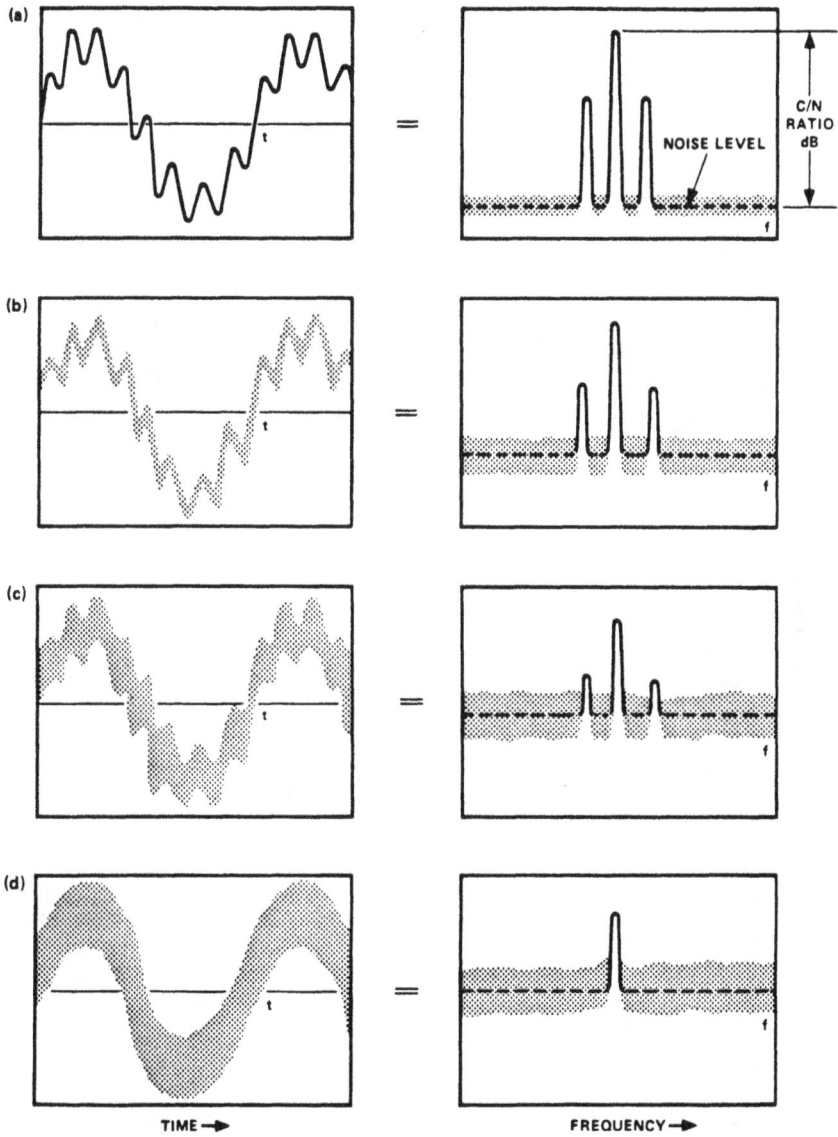

Figure 9.8 Modulation Being Obscured by Increasing Noise Shown in the Time Domain as well as the Frequency Domain. Note: The noise increases in the sequence (a), (b), (c). In (d) another noise level increase completely obscures the sidebands.

Source: Hewlett Packard, Cable Television System Measurement Handbook

Table 9.4
Noise Terminology

NAME	DESCRIPTION	DEFINITION
Noise Power Density, NP	Noise power level over a specific band of frequency in watts	NP (Watts referred to Hz) $= kTB$ k = Boltzmann's constant 1.374×10^{-23} T = Temperature in °Kelvin (Room Temp = 290° K) B = Frequency Bandwidth, Hertz NP (dbW) $= 10 \log_{10} kTB$ $\quad$ (dBm) $= 10 \log (kTB) + 30$ NP (dBmV) $= 10 \log (kTB) + 78.75$
Carrier to Noise Ratio, C/N	Power difference between carrier signal and noise power density, in dB	$C/N = C\text{-}N$ C = Carrier Level Power, dBmV NP = Noise Power Density, dBmV/unit bandwidth
Noise Figure, Noise Factor, NF	Input S/N to Output S/N of amplifier (where S/N = signal-to-noise ratio)	$NF = \dfrac{S/N \text{ Input}}{S/N \text{ Output}}$ = Noise Factor NF (dB) $= 10 \log \dfrac{S/N \text{ Input}}{S/N \text{ Output}}$ = Noise Figure where S/N is power ratio, not dB

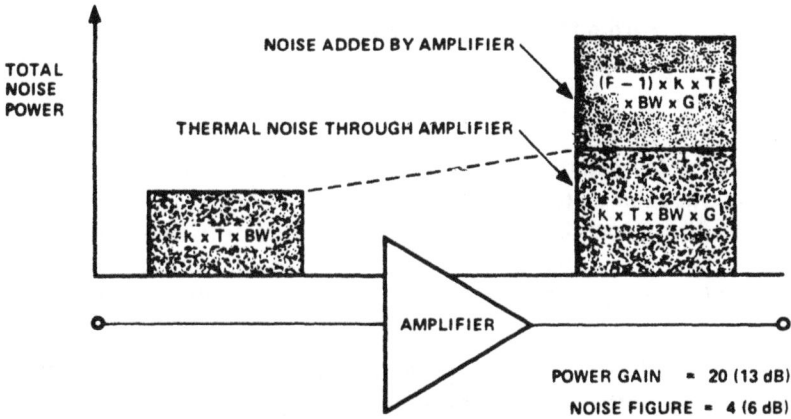

Figure 9.9 The effect on noise power by an amplifier. The input noise is amplified directly by the amplifier gain. An incremental amount of noise is added, depending upon the noise figure and gain of the amplifier.

Source: Hewlett Packard, Cable Television System Measurement Handbook

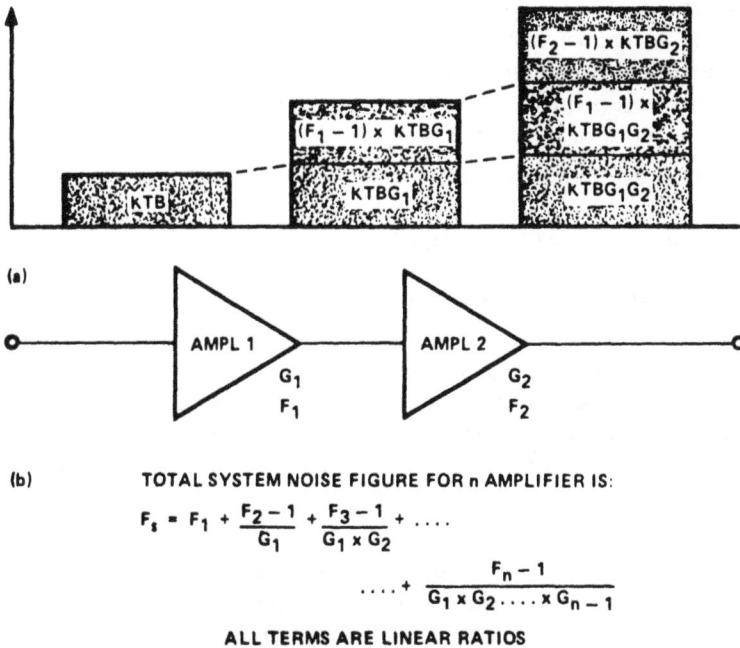

TOTAL SYSTEM NOISE FIGURE FOR n AMPLIFIER IS:

$$F_s = F_1 + \frac{F_2 - 1}{G_1} + \frac{F_3 - 1}{G_1 \times G_2} + \ldots$$

$$\ldots + \frac{F_n - 1}{G_1 \times G_2 \ldots \times G_{n-1}}$$

ALL TERMS ARE LINEAR RATIOS

Figure 9.10 System Noise Figure for Cascaded Amplifiers
Source: Hewlett Packard, Cable Television System Measurement Handbook

9.5 Interference from Outside the System

There are a number of entrances for unwanted signals in a CATV system. They can enter through the antenna at the head-end, or be picked up by leaky RF field distribution equipment. One common interference signal comes from a strong local station radiating onto the cable system. If the station is carried on the cable system in the same channel slot then the TV receiver will show a leading ghost picture, because of the system distribution delay. The ghost would appear as background to any channel put on the system at that broadcast frequency. The term *direct pick-up* is used to describe the type of interference.

When two broadcasting stations using the same channel are within pick-up distance of each other, their carriers are offset by ±10 KHz to prevent TV receiver interference. The CATV antenna picks up these "co-channels" along with the desired signals and distributes them unimpeded (because they are well within the channel passband filters), resulting in *co-channel interference*. If the co-channel level is high enough, the TV receiver will display two channels on one.

Figure 9.11 illustrates the appearance of co-channel interference in the frequency domain. At interference levels of –50 dB and above, picture distortion is evident. As shown, the audio of the channel is also distorted by the co-channel audio signal. However, the specification centers only on the video carrier because it is the reference for all the signals in the channel passband and contains most of the channel's energy.

There are a great number of other outside signals which can cause interference, both discrete and broadband, but generally CATV systems are sufficiently isolated so as not to be disturbed.

9.6 Interference from Within the System

More serious in nature are the interference signals which fall inside the TV spectral bandwidth that is generated within the CATV system itself. The simplest of these, hum, is an amplitude modulation of the carrier by a signal whose frequency is usually a harmonic of the power-line frequency. It can be generated from any number of the active devices or passive connectors along the distribution line. Figure 9.7 explains the fundamentals of how these are generated.

In the frequency domain, hum sidebands appear as two signals placed symmetrically on either side of the carrier and spaced the line frequency by an amount equal to or its harmonic, as Fig. 9.12 shows. One or two horizontal bands appear on the TV picture when interference levels exceed –32 dB down; that is, hum sidebands are less than 32 dB from the carrier peak. For a 60 Hz

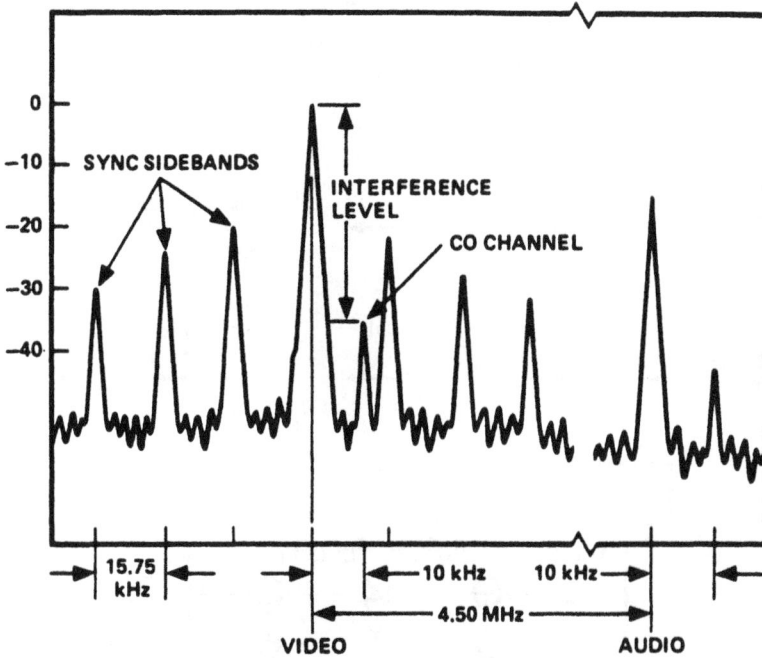

Figure 9.11 Co-channel Interference on the Video and Audio of a Single Channel in the Frequency Domain. Note: The specification is measured from the video carrier peak to the top of the interfering carrier.

Source: Hewlett Packard, Cable Television System Measurement Handbook

monochrome TV transmission, these lines will be stationary and, for the 59.94 Hz rate of color transmission, the 60 Hz hum lines will slowly move through the picture in the opposite direction of the field scan.

The other major internal system interference is intermodulation. Intermodulation products are extra signals caused by non-linearities of active components such as amplifiers. Figure 9.13 gives a simple graphic example of intermodulation products from two signals.

An intermodulation product's frequency is given by

$$n_1 f_1 \pm n_2 f_2 \pm n_3 f_3 \ldots \infty$$

where

f = the frequency of a system carrier pilot or other signal
n = the harmonic number of f

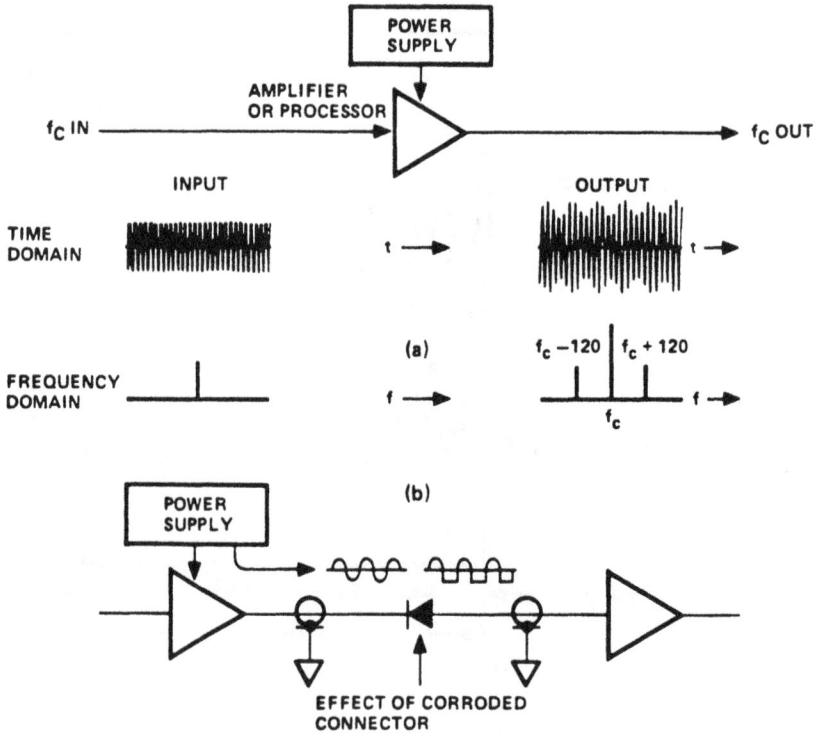

Figure 9.12 Hum Sideband Generation: (a) An active device such as distribution amplifier or processor modulates the RF wave. (b) When ac power is fed down the cable to power several amplifiers a corroded connector can act like a crystal diode, partially half-wave rectifying the 30 or 60 volt ac wave.

Source: Hewlett Packard, Cable Television System Measurement Handbook

Intermodulation products are given an order number for the sum of n in the frequency formula. For example:

$1f_1 + 2f_2$ is a $(1 + 2)$ or *3rd Order intermod product;*

$2f_6 - 3f_1$ is a $(2 + 3)$ or *5th Order intermod product;*

$f_1 + f_2 - f_4$ is a $(1 + 1 + 1)$ or *3rd Order intermod product.*

Generally, low-order intermodulation are the strongest intermodulations. The order number takes into account both the harmonic content and the number of mixing steps in the intermodulation. It is natural for higher

AMPLIFIER OR
OTHER ACTIVE CIRCUIT

(a) MIXING   INPUT

f_1 f_2

OUTPUT

$f_2 - f_1$ $f_1 + f_2$

(b) HARMONICS

f_1 f_2

f_1 f_2 $2f_1$ $3f_1$
$2f_2$

(c) COMBINED (a) AND (b)

f_1 f_2

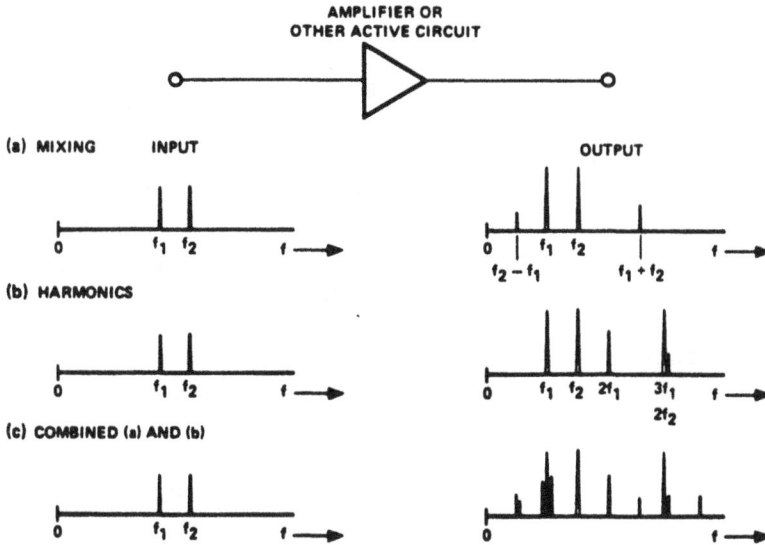

Figure 9.13 The Source Intermodulation Products: (a) A number of active and passive components in the distribution system form additional outputs from the input signals. (b) Other undesirable outputs occur when an amplifier with a nonlinear response produces harmonics of the input frequency. (c) If combined, these mixing and multiplying faults of the distribution components can cause a multitude of responses, called intermod products.

Source: Hewlett Packard, Cable Television System Measurement Handbook

harmonic signals (higher n) to be lower in amplitude level. Also, the product of a mixing process is lower than either of the mixed signals, so each mixing step reduces the signal level. The highest level third-order intermodulation product is triple beat (see Ch. 8). Triple beat (TB) is an $f_1 \pm f_2 \pm f_3$ product. Because each signal is a fundamental ($n = 1$), the power available for mixing is high. The only power lost is in the two mixing processes $f_1 \pm f_2$ and $(\pm f_1 \pm f_2) \pm f_3$. The next most troublesome intermodulations are the third-order $f_1 \pm 2f_2$ products.

By the simple rule above, second-order products should be the most powerful, as they would be were it not for the fact that distribution equipment is designed to specifically minimize second-order distortion products.

When intermodulation products fall in the channel passbands, interference occurs. There is no obvious rule or pattern for the places in the channel passband where these can fall, as opposed to hum and co-channel interference. Therefore, the subjective (or TV receiver) response cannot be predicted. Because the make-up and frequency of the intermodulation disturbances are essentially unpredictable in nature, system specifications are tougher. A −57 dB signal will not show subjective interference. This is 10 dB (i.e., a factor of 10 times less power) more stringent than co-channel interference and 14 dB (i.e., 25 times less power) more stringent than hum.

The number of intermodulation products possible in a system is surprisingly high, as Table 9.3 demonstrates. The table gives a partial list of the triple beat ($f_1 + f_2 - f_2$) products for a 12-channel, 2-pilot system that fall within the high and low bands.

Crosstalk in stereo sound reproduction and telephone voice communication has its counterpart in CATV systems. This specific type of intermodulation is called cross-modulation (X-Mod). (See Ch. 8.) Cross-modulation is particularly troublesome in TV spectra because of the type and amount of modulation carried by the TV signal.

Table 9.5

Triple Beat Intermodulation Product Example

System Signals		Possible $f_1 \pm f_2 \pm f_3$ Products Falling Within:			
Channel	MHz	53 to 88 MHz		174 to 186 MHz	
2	55.25	53.25 MHz	71 MHz	175 MHz	195.25 MHz
3	61.25	55.0	71.25	176.25	195.5
4	67.25	55.5	73.25	177.25	197
5	77.25	57.25	77	177.5	197.25
6	83.25	57.5	79.25	179	199
7	175.25	59	79.5	181	199.5
8	181.25	59.25	79.75	181.5	201.25
9	187.25	61	81	183.25	201.5
10	193.25	61.5	82.25	183.5	203
11	199.25	62.5	83.5	185	203.25
12	205.25	63.75	85.25	185.25	205
13	211.25	65	85.5	187	205.5
Pilot I	73.5	65.25	85.75	187.5	207.5
Pilot II	118.25	67.5	87	189.25	209
		69.75	87.25	189.5	209.25
				191	210.25
				191.25	211.25
				193	211.5
				193.5	

Cross-modulation, like crosstalk, simply means a desired channel is being modulated by another; that is, some of the modulation sidebands on the desired channel are caused by another channel. In telephone crosstalk, this effect is a second conversation on the line while you are trying to talk. In television, the effect is jittery diagonal stripes on the picture, generated by the 15.75 synchronized pulses of other channels being impressed upon the received channel.

Before tackling some of the specifics of cross-modulation, let us look closely at the TV spectral response once more. Figure 9.14 shows a high resolution frequency domain display of the TV video carrier. The most dramatic characteristics are the evenly spaced sidebands. This spacing is the signal's modulation rate of 15.75 kHz. The amplitude characteristics, however, can cause confusion about the measurement standards for cross-modulation.

To understand the amplitude characteristics, it is necessary to briefly review the spectral characteristics of amplitude modulation in Fig. 9.8. As the amplitude modulation (AM) index is increased, the carrier remains constant and the sidebands increase. Thus, the total signal power increases. This is not the case with the TV carrier modulation. However, it is similar enough to an AM scenario insofar as its sideband energy can be used to show a relation to modulation level and picture quality.

Given a clear, unmodulated channel, let us say the processor standby of channel 4, any extraneous 15.25 kHz sideband in the system could be detected. This sideband content would be the amount of modulation imposed upon

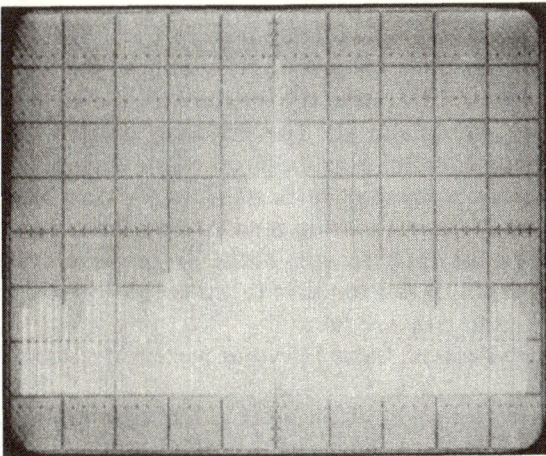

Figure 9.14 Spectrum Analyzer Display of Carrier and 15.75 kHz Sidebands Shows the Complexity of Video Modulation.
Source: Hewlett Packard, Cable Television System Measurement Handbook

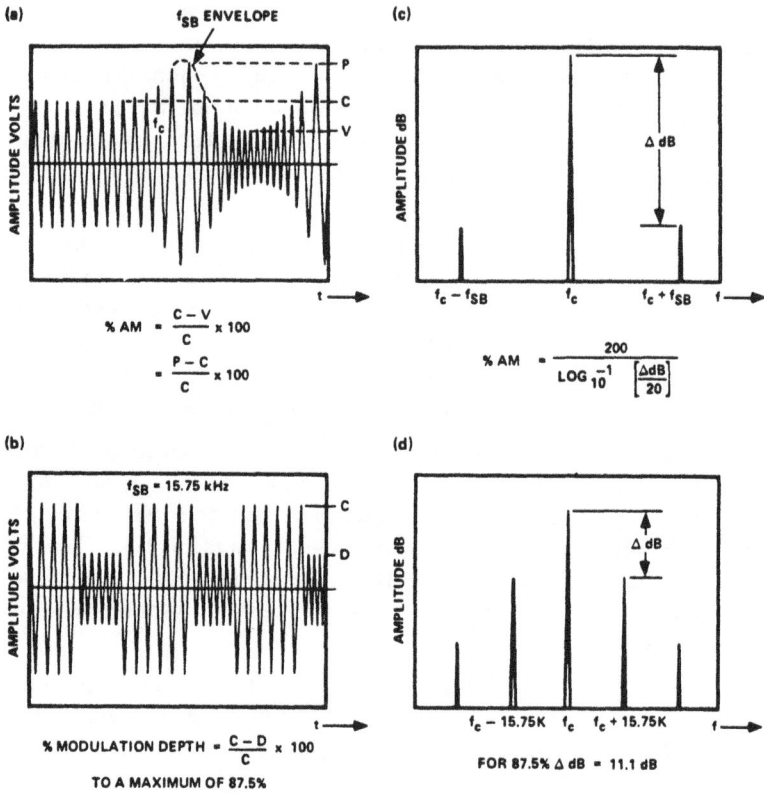

Figure 9.15 Typical AM and its Relation to TV Carrier Modulation. (a) and (b): The time domain displays of both the typical AM and a simplified TV carrier showing how their levels of modulation are defined. (c) and (d): The frequency domain displays of each modulation showing the direct relationship of sideband level to percent modulation in the standard AM case. Note: With the TV carrier the relationship cannot be put into an equation because the pulse characteristics of the TV carrier add sideband energy which does not conform to either the AM nor the pulsed RF spectral response rules.

Source: Hewlett Packard, Cable Television System Measurement Handbook

the unmodulated channel by all the other system channels operating in their normal fashion. The only valid measurement is comparison of the sideband power with the carrier power, as in AM. However, TV is different from the AM and pulse modulation cases, neither, and cross-modulation cannot be quantified in the same terms (percent modulation or desensitization).

When regular programming is restored to the test channel, the cross-modulation sidebands will not be visible in the frequency domain. They may still be observed subjectively, but they will be buried under the regular modulation.

The cross-modulation level is the power difference (ratio), between the unmodulated carrier and the 15.75 kHz sideband. Picture interference may begin to appear at –46 dB cross-modulation level. It shows up as stripes across the screen, sometimes moving in rapid succession in a so-called "waterfall" effect, or, depending upon the phasing of the interfering signals, a wider and slower band known as a "wiper" effect. The wiper effect is usually the more objectionable.

9.7 Signal Leakage

Ideally, all the distributed CATV signals are contained within the cable network, but signal leakage from cable systems can radiate causing interference with local VHF/UHF communication. Air navigation receivers could be misled by the radiated CATV intermodulation beat frequencies which can escape from a leaky distribution system. Corroded connectors and other distribution hardware break-down are generally responsible for leakage, but high subscriber signal level can also be at fault. High signal levels at a customer drop can cause radiation through the subscriber's old TV antenna, if it is still connected to the receiver.

The major causes of distribution systems radiation are leakage and poor ground, both the result of hardware failure. Signal radiation by leakage means the CATV signals are radiating from an opening in the shielding of the distribution system, i.e., the opening is acting like an antenna. Poor grounding makes even leakproof shielding ineffective. An undergrounded outer conductor simply re-radiates the center conductor's signal to the outside.

Let us look at the characteristics of the radiated signals. Radiation of a signal presupposes the transfer of energy and, therefore, the flow of current. Current induces a magnetic field which can be used to quantify the amount of energy in a radiated signal. Space has an associated impedance of 377 Ω which converts current flow into a voltage differential E over a section of space. The relationship is

$$E/H = 377\,\Omega \qquad\qquad\qquad (9\text{-}7)$$

where

E = electric field strength, rms volts per meter

H = magnetic field strength, webers per square meter or tesla

This defines field strength. So, by knowing either E or H, you know the other.

An antenna translates the spatial field strength into volts, an easily measured quantity. At any one frequency, the field strength blanketing the antenna will produce a specific voltage at the antenna terminals. The relationship between this voltage and the field strength is the antenna correction factor K. From Fig. 9.16, we see that

$$E = KV \tag{9-8}$$

A convenient unit for V is dBμV. Thus, if we convert this equation to dB referenced to a microvolt,

$$E \left(\frac{\text{dB} \mu \ V}{\text{meter}} \right) = K \left(\frac{\text{dB}}{\text{meter}} \right) + V \left(\text{dB} \mu \ V \right) \tag{9-9}$$

The manufacturer's specifications for an antenna will include a table that shows the antenna correction factor, K, *versus* frequency in decibels/meters (dBm). Then the E field is equal to the antenna voltage V, in dBV, plus the correction factor K, in dBm. It is convenient to graph the correction factor and on the same graph plot the E field specification limit from the radiation test of interest. Plotting the difference of these, $E - K$, results in plot of the maximum allowable level of receiver voltage for this particular radiation test. It is now only necessary to plot test receiver voltage on the same graph to determine whether the radiation exceeds the specification level; no conversion to dBV/m is necessary. Figure 9.17 shows an example of test levels for a particular radiation specification.

Some antenna manufacturers characterize their antennas with isotropic power gain *versus* frequency. Antenna factors can be derived from gain using

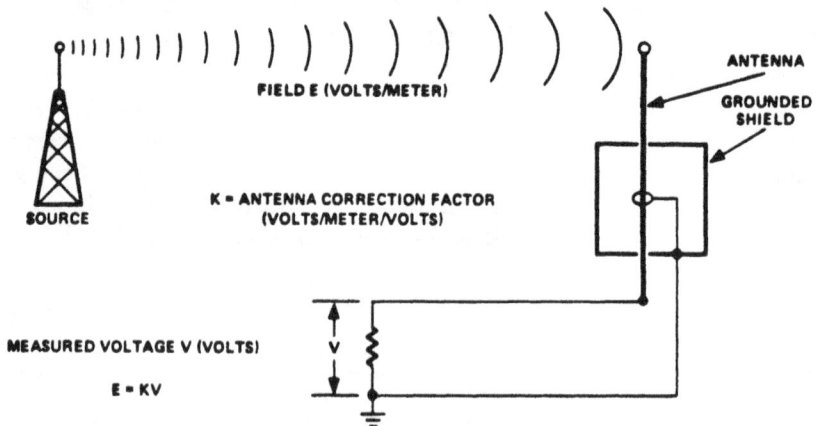

Figure 9.16 Radiated Measurement Terms
Source: Hewlett Packard, Cable Television System Measurement Handbook

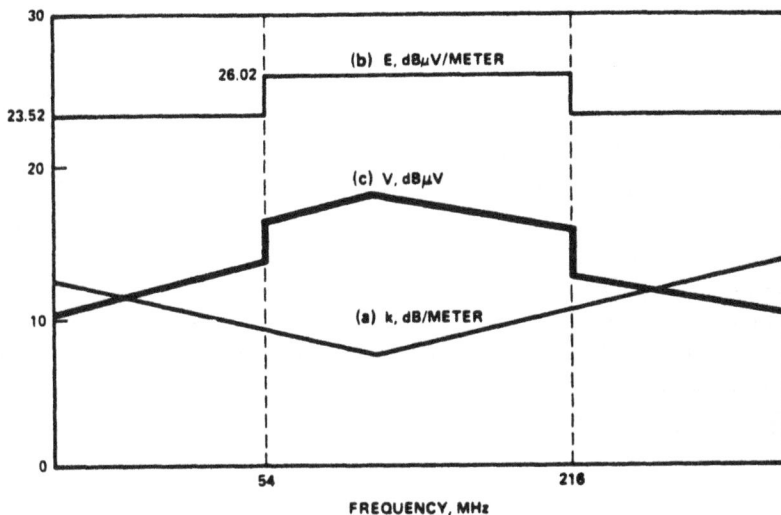

Figure 9.17 Test Levels. Note: Test levels for a particular radiation specification can be put into units of receiver voltage by ploting (a) the antenna correction factor in dB/meter,(b) the test limits, and (c) the difference.

Source: Hewlett Packard, Cable Television System Measurement Handbook

$$K = 20 \log f - 10 \log G - 12.79 - 10 \log Z_o$$

where

    K = antenna factor, decibels/meters, dBm   (9-10)

    f = frequency, MHz

    G = antenna gain, unitless power ratio

    Z_o = receiver input impedance, ohms

For 75 Ω input impedance

$$K = 20 \log f - 10 \log G - 31.54 \qquad\qquad (9-11)$$

Receiver bandwidths must be as wide as signal level measurement bandwidths because radiated measurements are expressed as rms peak of the synchronizing pulse.

Measuring radiated signals in the presence of strong broadcast signals at the same frequency is not possible. Only signals unique to the CATV distribution system can be used to evaluate the system's radiation. There are usually a number of pilot tones and translated TV channels that can be used to characterize the system.

9.8 Summary

Network analysis measurements characterize a circuit's behavior with a given signal. Signal analysis measurements define a signal, regardless of its source. Therefore, CATV system behavior is characterized more as a signal than a network measurement.

Signals are characterized by frequency (accuracy and stability), level (voltage, power, the logarithmic form of power, dB, and flatness), and noise (density, bandwidth, and signal to noise ratio). CATV signals are defined and specified in terms of frequency (carrier accuracy), level (adjacent carriers, channel frequency response and absolute power), and noise (carrier-to-noise ratio). Extra signals in the CATV spectrum are interference (co-channel, hum, cross-modulation, and intermodulation). Radiation is interference outside the system by signal leakage.

10 Test Instrumentation

Now that the various measurement parameters have been reviewed, let us summarize the many types of equipment that can be used for CATV system performance testing and outline the strengths and weaknesses of each.

10.1 Network and Signal Analysis

As mentioned in the previous chapter, types of electronic measurements generally fall into two categories, network analysis and signal analysis. Network analysis involves measurements which display how a known signal is changed by a specific circuit. In CATV measurements both techniques are useful, but signal analysis is the most useful in CATV system operation measurements.

10.2 Signal Analyzers

Signal analysis instruments come in a great many forms. Here is a familiar list:

- frequency counter
- oscilloscope
- power meter
- ac volt meter
- field strength meter
- wave analyzer
- spectrum analyzer
- spectrum viewer
- noise figure meter

Each is capable of measuring one or more of the parameters on an amplitude *versus* frequency plot. Frequency can be measured by all but the noise meter, ac volt meter, and power meter, while power amplitude can be measured on all but the frequency counter.

To understand how each of these instruments makes its specific signal analysis measurement, let us use the circuit and signal spectrum of Fig. 10.1.

The *frequency counter* measures and displays the frequency of the signal at its input. It does this by counting the number of cycles the input signal goes through during a very accurately timed interval. The number of cycles is then scaled to display the frequency in hertz. (See Fig. 10.2.) At test points 1 and 2 of Fig. 10.1, the counter will read 30 MHz and 5 MHz, respectively, as long as the signal level is high enough to be counted. At point 3, however, the three higher level frequencies present on the cable, 35 MHz, 25 MHz and 30 MHz, would cause extra counts in the counter's timed interval resulting in a false reading. Similar problems occur if the signal is modulated as in test point 5.

Counters are capable of more than sufficient accuracy for CATV frequency measurement. Tapping off converter oscillator frequencies directly before mixing, or using a tunable bandpass filter in front of the counter, will ensure accurate readings. When measuring modulated carrier frequencies, the bandpass filter must be narrow enough to exclude the first 15 kHz sidebands. An FM audio carrier would require a wider bandpass filter and long counter gate times to average out the effects of the FM.

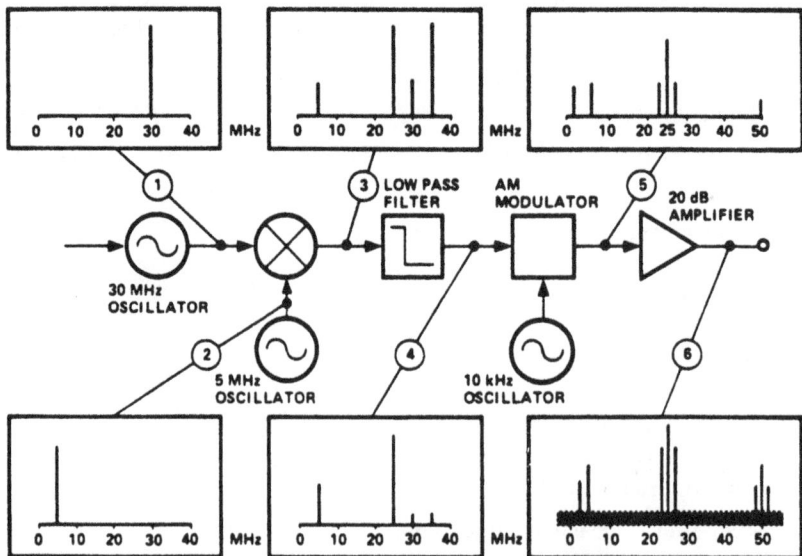

Figure 10.1 Sample Circuit with Test Points and Resultant Signal Spectrum
Source: Hewlett Packard, Cable Television System Measurement Handbook

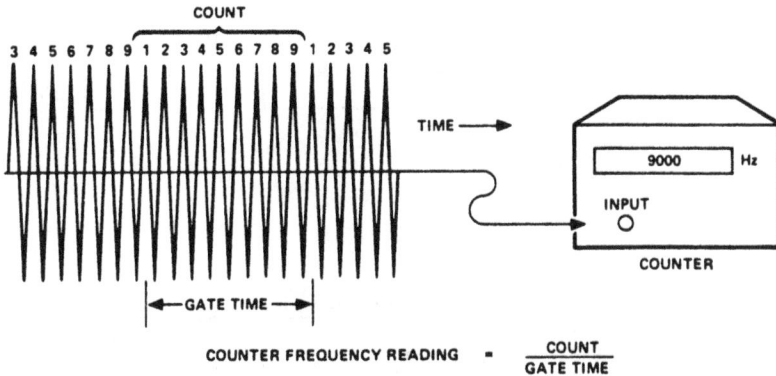

COUNTER FREQUENCY READING = $\dfrac{\text{COUNT}}{\text{GATE TIME}}$

Figure 10.2 Frequency counter principle. Note: The counter counts the number of cycles in the input signal for a specific period of time, called gate time. Logic circuitry then translates this information ito frequency.

Source: Hewlett Packard, Cable Television System Measurement Handbook

The *oscilloscope*, which provides a video display of the actual signal voltage waveform in the time domain, is mostly used in CATV as a waveform monitor to inspect the TV carrier modulation envelope. All signals at the input are presented simultaneously on the display. Thus, it is difficult to discern any individual signal's voltage or frequency.

The *ac volt meter* and *power meter* measure the rms voltage and power, respectively, of a signal input. Except for the frequency limitations imposed on the instruments themselves, neither the volt meter nor the power meter is frequency selective. If the voltage or power at test points 5 of Fig. 10.1 were measured, the reading would lump the power of all the frequency components together.

In CATV, the ac volt meter and power meter are valuable where total voltage and power measurements are needed (e.g., transmitter power output), and amplitude measurements of single frequency sources can be very accurately made whether modulated or not.

If the signal into an ac volt meter or power meter were somehow limited to a narrow bandwidth somewhere in the circuit, specific signal voltages and power could be measured. Such an instrument is called a *tuned volt meter*. "Tuned" simply means the amplitude measurement has been made frequency sensitive. The *field strength meter* and the *wave analyzer* are tuned volt meters (their operating principles are reviewed in Fig. 10.3). Other names for these instruments are *signal level meter* and *wave meter*, respectively.

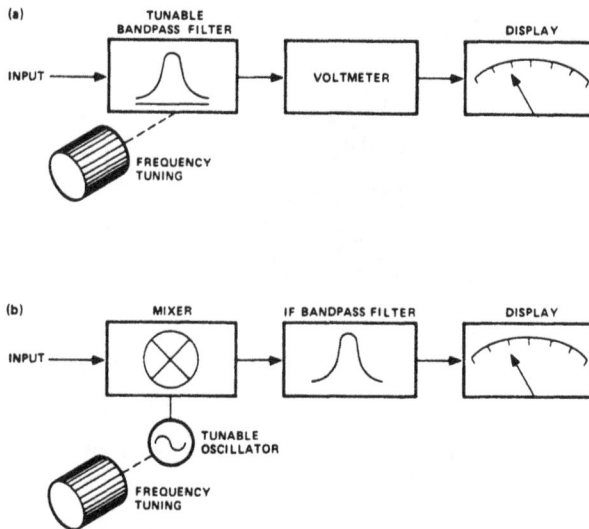

Figure 10.3 Tuned AC Volt meter, Field Strength Meter and Wave Analyzer. Note: Two design approaches are used to make an AC volt
meter frequency sensitive (tunable): (a) A tunable bandpass filter
is placed in front of a broadband AC volt meter; (b) Input signals
are mixed with a reference oscillator to provide an intermediate
frequency (IF) signal which is then measured (the heterodyne
process). All the bandpass filtering can be done at IF
frequencies.

Source: Hewlett Packard, Cable Television System Measurement Handbook

The display for these instruments is usually a meter amplitude and a dial or
digital readout for frequency. Thus, the instrument measures a voltage or
power for each frequency in its range. The spectrum of a signal can be
characterized by taking point-by-point readings and graphing the data.

The tuned volt meter is capable of making accurate CATV level instruments.
Its most valuable asset in this respect is the narrow bandwidths it can achieve.
The narrow bandwidth is the "window" through which the analyzer *sees* a
signal. The narrower this bandwidth is the better it resolves close-in sidebands
such as hum. Figure 10.4 shows how bandwidth figures resolve power.

The wave analyzer has sufficient resolving power but lacks the frequency
range to cover all the TV carriers. This resolving power is not lost, however,
because it can be used to view the 6 MHz baseband of processors and
converters or the IF of many broadband heterodyne tuned receivers.

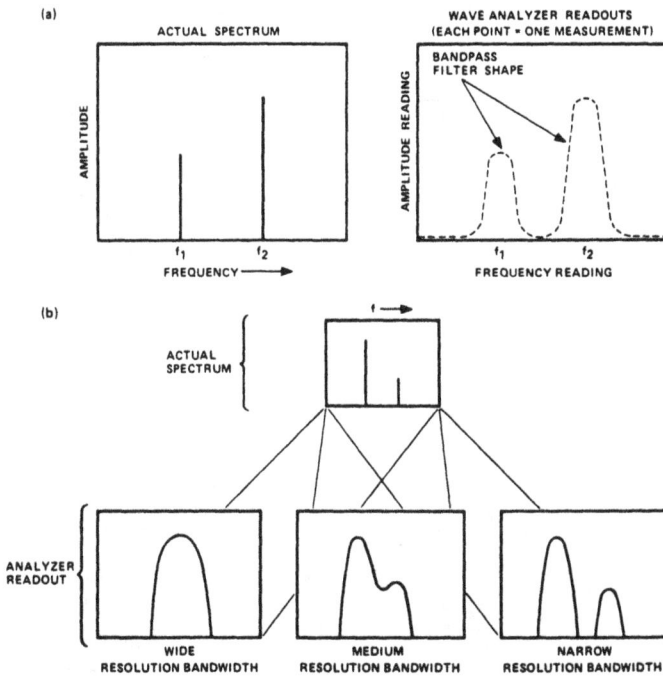

Figure 10.4 Bandwidth and Resolving Power of Wave Analyzers: (a) Shows how the analyzer perceives signals in a specturm; (b) The effect of progressively narrower bandwidths on the same signal. Note: The narrower the filter the closer the output display resembles the actual spectrum.

Source: Hewlett Packard, Cable Television System Measurement Handbook

For measurement simplicity, the frequency of the tuned volt meter can be automatically swept so that a continuous readout can be displayed on an amplitude *versus* frequency plot. This so-called *swept-tuned volt meter* results in the *spectrum viewer* and *spectrum analyzer*. (See Fig. 10.5.)

The same resolution rules apply to swept-tuned apparatus as to the manually tuned volt meters, except that the sweep time must be slow enough to allow the filter components adequate response time.

The spectral displays provided by these instruments are ideal for CATV because they show more about the signal than any other technique. Signal

(a) SPECTRUM VIEWER

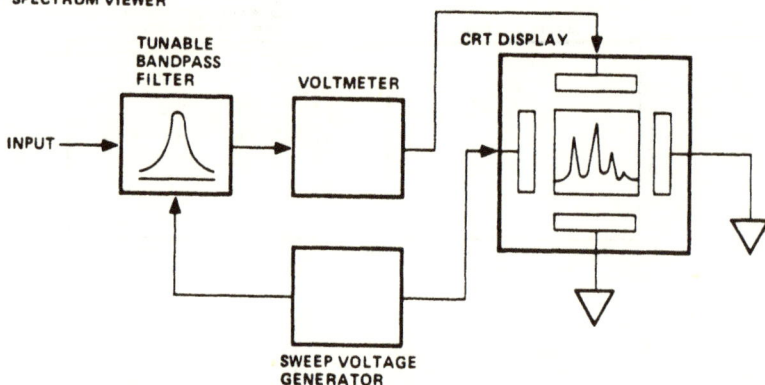

(b) SPECTRUM ANALYZER

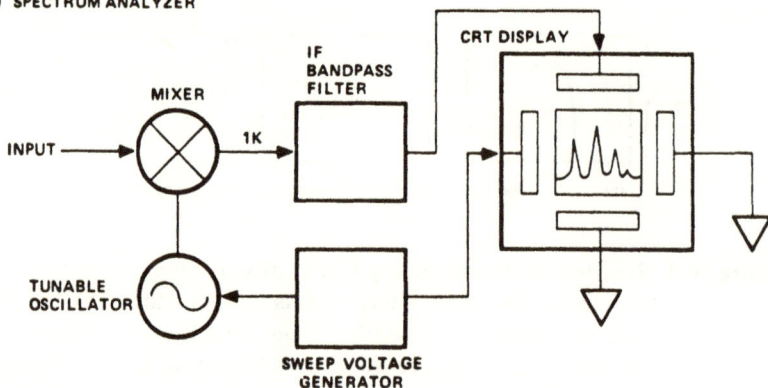

Figure 10.5 (a) Spectrum Viewer (b) Spectrum Analyzer
Source: Hewlett Packard, Cable Television System Measurement Handbook

level, noise, sidebands, and interference can all be measured. The spectral display gives instant insight into the operation of the system and its components. The spectrum analyzer can be used as a fixed-tuned receiver and as a wave analyzer used on the video output for detailed signal sideband analysis. The spectrum analyzer, with a companion signal source called a *tracking generator*, can measure the frequencies of very low level signals, even in the presence of high level signals, with frequency counter accuracy.

The *noise figure meter* is used to determine noise figure (NF), not noise level. It measures the noise contribution made by an amplifier rather than the noise being generated from a CATV system. The *noise meter*, also known as strength meter, makes absolute noise power (NP) measurements. It is a specialized test instrument generally used to make radio frequency interference (RFI) measurements to ensure electromagnetic compatibility between different pieces of equipment.

10.3 Summary

Most of the commonly used CATV test instruments are best defined as signal analyzers. The counter measures the frequency of a single frequency signal. The oscilloscope (waveform monitor) provides an amplitude (voltage) *versus* time display of a signal. The ac volt meter and power meter measure signal level over a wide bandwidth. The tuned volt meter (wave analyzer and field strength meter) measures signal level at a specific frequency. The spectrum analyzer is an automatic tuned volt meter, capable of displaying a continuous spectrum and a *noise meter* is used to determine absolute noise power density measurements.

11 Characteristics of Coaxial Cable

In dc circuits we normally think of the connecting wires and lines as ideal, without resistance or reactance. With ac signals, when the length of a line is more than one-tenth of a wavelength at the highest frequency of interest, the properties of the line must be considered. Each of the conductors in a coaxial cable has a respectable length as compared with a wavelength, creating a significant inductance. The conductors are insulated from each other, but are in close arrangement, which will probably produce some capacitance. The actual inductance and capacitance is distributed uniformly along the line. There are two properties of capacitances and inductances that are used in studying transmission lines:

1. *The voltage across a capacitor cannot change instantly.* The voltage depends on a charge building up. If a capacitor is suddenly switched into a circuit, it will appear as a short at first until it has time to take a charge.

2. *The current in an inductance cannot charge instantly.* Any change in current is opposed by the counter electromotive force (EMF) produced by the magnetic field associated with the inductance. If an inductance is suddenly switched into a circuit, it will look like an open circuit for a moment. The characteristic impedance depends on the value of L and C in a circuit, and these in turn depend on the dimension of the line. The value is given by the equation:

$$Z_0 = \sqrt{L \cdot C} \qquad (11\text{-}1)$$

where

Z_0 = characteristic impedance in ohms

L = inductance

C = capacitance

Since the inductance and capacitance are distributed along the line, they are given in units such as henrys and farads.

11.1 Attenuation

Attenuation is the paramount problem in CATV systems. No wire is a perfect conductor. Therefore, as signals travel through coaxial cables, they become weaker and weaker. This weakening of the signal is known as *attenuation*.

The longer a CATV cable, the more it attenuates the signal. Therefore, attenuation is usually stated in terms of dB per hundred feet.

Characteristic impedance is almost independent of frequency, but attenuation is not. Attenuation is directly proportional to frequency. In other words, the higher the frequency, the more the cable will attenuate the signal. A given cable will generally triple the dB measure of loss at channel W as it causes at channel 2.

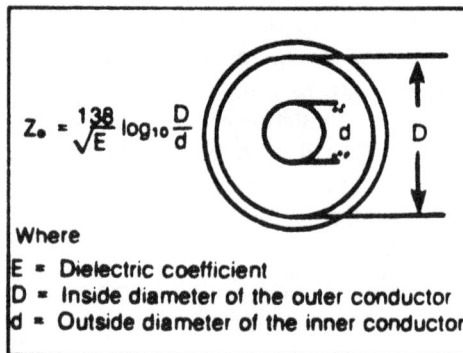

$$Z_\circ = \sqrt{\frac{138}{E}} \log_{10} \frac{D}{d}$$

Where
E = Dielectric coefficient
D = Inside diameter of the outer conductor
d = Outside diameter of the inner conductor

Figure 11.1 Attenuation
Source: Texscan Corp.

Thus, if you increase *d* (the diameter of the center conductor) without also increasing *D* (the diameter of the shield) or changing *E* (the dielectric constant), the characteristic impedance will change. Therefore, the only way to use a large center conductor is by making the overall cable larger, or using a different dielectric.

There is a practical limit to the size of cable that can be handled. Most systems use shield outer diameters (abbreviated OD) of 0.412″, 0.500″, or 0.750″ in their trunk and feeder lines.

As the formula indicates, one way to reduce attenuation is to use a dielectric with a lower dielectric constant. At present, this is the only practical way, so most recent improvements in CATV cables involve the use of different dielectric materials.

The earliest CATV cables used solid polyethylene as a dielectric. Later, it was found that polyethylene could be foamed to form tiny air spaces. Because cables made with foam polyethylene cause little more than half as much attenuation as solid polyethylene cables, foam quickly gained acceptance. However, solid polyethylene is more moisture resistant than foam, and so it is often preferred for underground construction.

After foam polyethylene, the next step in lowering cable attenuation was the use of polystyrene as a dielectric. Polystyrene is a light, brittle substance, filled with air holes. Polystyrene lowers attenuation by about 20%, compared with foam polyethylene. The disadvantages of polystyrene are mechanical. It is relatively hard to work with and it collapses more easily. Thus, polystyrene cables are more likely to be damaged during plant construction, unless they are handled with care.

Air dielectric cables have been available for CATV use for many years. Air dielectric has the lowest dielectric constant of any of the types discussed herein. However, disks of some sort are required to support the shield away from the center conductor, so that air dielectric cables tend to have mechanical disadvantages making them relatively unpopular.

As the formula indicates, one way to reduce attenuation is to use a dielectric with a low dielectric constant. The dielectric constant of air dielectric is 1.0, foam polyethylene is 1.5, and solid polyethylene is 2.3. Therefore, cables using foam dielectrics cause less attenuation than similar cables using solid polyethylene dielectric.

11.2 VSWR

To understand why, let us examine the concept of voltage standing wave ratio (*VSWRC1*). *What does VSWR* really mean? When we discussed characteristic impedance, we said that maximum signal transfer occurs when the generator is matched to the transmission line. During that discussion, we assumed that the cable was a perfect 75 ohm impedance at all points and that it was correctly terminated by 75 ohms. What if this is not so?

Take for example, a 75 ohm cable terminated by a 25 ohm load. Obviously, maximum signal transfer cannot occur in this case. Only part of the signal is absorbed by the 25 ohm load. What happens to the signal that is not absorbed by the load?

Remember that once the signal is half a wavelength away from the generator, it is on its own. It moves the line as a series of alternate build-ups and collapses of magnetic fields which induce current. This package of energy is called an *incident wave*.

When the incident wave hits the load, the part that is not absorbed has no place to go. So, it simply turns around and goes the other way, by the same process of collapsing magnetic fields inducing current. For obvious reasons, the part of the signal that bounces back is called the *reflected wave*.

At this point, we have the problem of current flowing through a single conductor in two directions simultaneously. In reality, two opposite currents cannot flow in the same portion of the same wire at the same time, but two opposite fields can coexist in the same place. Therefore, the two fields are passing through each other.

At some points along the coaxial line, the polarity of these passing fields is such that they add together. Thus, the current is increased at these points. At other points, the two fields are of opposite polarity and, thus, subtract from each other. Where the two fields are equal and opposite in polarity, no current flows.

Both the incident and the reflected wave are moving, but they always cross each other at exactly the same points. Therefore, the points of addition and cancellation on the center conductor are always the same and the resultant wave pattern is stationary. Because they do not move, we call these *standing waves*. Because standing waves cause ghosts and color smears, they are highly undesirable.

It ought to be easy enough to eliminate standing waves by simply terminating all cables correctly. What if there is a fault in the cable itself? For example, what if the center conductor of the cable is drawn too fine at one point, or the shield is dented which changes the distance between the shield and the center conductor? These and other imperfections may not be visible to the eye, but they cause a mismatch in the cable at the points where they occur. These mismatch points, in turn, cause reflected signals which result in standing waves. The worse the mismatch, the greater the amplitude of the standing wave.

The *VSWR* is the ratio between the maximum voltage and the minimum voltage along the line. The formula is

$$VSWR = \frac{E_{max}}{E_{min}} \qquad\qquad (11\text{-}2)$$

The best possible condition would be where E_{max}, and E_{min} are equal. In this case, there would obviously be no standing waves and the *VSWR* would be equal to 1. Let us assume that E_{max} and E_{min} are both equal to 1.7. Then

$$VSWR = \frac{E_{max}}{E_{min}} = \frac{1.7}{1.7} = 1 \qquad\qquad (11\text{-}3)$$

Now suppose that E_{max} is 2.0 and E_{min} is 1.4.

$$VSWR = \frac{E_{max}}{E_{min}} = \frac{2.0}{1.4} = 1.43 \tag{11-4}$$

$VSWR$ is often written as a ratio, e.g., 1.5 to 1, or, more commonly, 1.5:1. The lower the $VSWR$, the better the cable or other device. No unit in a CATV system should have a $VSWR$ greater than 1.5, and a good cable will have a $VSWR$ of about 1.1.

Direct measurement of $VSWR$, which cannot accurately be made under 1.5, is not practical for testing CATV cable quality.

11.3 Structural Return Loss (SRL)

A much more sensitive method for determining the cable quality involves the measurement of *structural return loss* (SRL). The return loss of a cable is the difference in dB between a signal applied to a properly terminated cable and the signal reflecting back through the terminal end.

The term *structural return loss* is commonly used for cable specifications because reflections within the cable are due to imperfections in the "structure" or physical characteristics of the cable. When referring to lumped constant devices, such as amplifiers or directional couplers, the term *return loss* is used.

The most common cause of structural imperfections in coaxial cable is a problem in the manufacturing process, resulting in small variations in cable dimension, which are repeated at equal intervals. This is known as *periodicity*. The resultant small reflections can accumulate to produce severe effects at certain frequencies. Great care is required in the production of such cable to avoid the problem.

The spacing between discontinuities in a length of cable determines the frequencies which will have the most reflected signal and the amplitude of the reflected signal. A small amount of signal will be reflected by the first discontinuity. Signal will also be reflected by the second discontinuity. These two reflected portions of the same original signal will travel together in the portion of cable between the source and the first discontinuity. At certain frequencies they will be in phase and thus will add together. At other frequencies they will be out of phase and will tend to cancel each other. This effect is multiplied by the number of discontinuities in that particular length of cable. If there are several, equally spaced discontinuities, these reflections will all be in phase at one or more frequencies and the structural return loss will be quite low.

11.4 Effects of Temperature Change on Cable Characteristics

The major system changes that occur with temperature are because of the changes in coaxial cable characteristics. The effect of greatest concern is the direct relationship between temperature change and cable attenuation. Because these factors are directly related, increases in temperature result in corresponding decreases in cable loss. If these changes remain uncompensated, hot weather would result in lower signal levels and more noise — especially at the system extremities — and cold weather would result in higher signal level and worse cross modulation.

Within the CATV industry, there is a basic rule-of-thumb to qualify the effects of temperature on cable attenuation. The rule-of-thumb states that a temperature variation of 10° F will preclude a 1% change in cable attenuation in all frequencies. This represents a general relationship derived from the more exact 0.11% per degree. It is interesting to note that temperature has exactly the same effect on coaxial cable as physically changing its electrical length. In fact, changing the electrical length of cable by +7.4% and -12.6% from the length at 73° F(23°C) corresponds to a temperature change over the range of +140° F to -40° F, respectively.

The impact of temperature change on system operation can be clearly demonstrated through the following example. Based on system design data, the average electrical spacing between amplifiers is 20 dB. Because the typical operational gain of a trunk amplifier is 22 dB, the 20 dB average spacing occurs at the typical design temperature of 68°F. If the system is designed for 35 channels, then the highest channel is W at a frequency of 295.25 MHz. Channel 2 is, of course, the lowest channel at a frequency of 55.25 MHz. Because cable length is always stated in dB of attenuation at the highest frequency, the spacing between amplifiers is 20 dB at channel W. The spacing of amplifiers at channel 2 is then

$$A_L = A_H \sqrt{\frac{F_L}{F_H}}$$

$$A_L = 20 \text{ dB} \sqrt{\frac{55.25}{295.25}}$$

$$A_L = 8.65 \text{ dB} \tag{11-5}$$

If the temperature now increases by 50° F to 128° F, the cable attenuation at all frequencies will increase by 5%.

$$\frac{1\%}{10° \text{F}} \times 50° \text{F} = 5\%$$

Therefore, the new spacing at 128°F is

Channel W 20 dB + (20 dB × 0.05) = 21 dB

Channel 2 8.65 dB + (8.65 dB x 0.05) =   9 dB

If the temperature decreases by 50°F to 18°F, the cable attenuation at all frequencies will decrease by 5%. Therefore, the new spacing at 18°F is

Channel W 20 dB   – (20 dB   x 0.05) =   19 dB

Channel 2 8.65 dB – (8.65 dB x 0.05) = 8.2 dB

A temperature change of 100°F has, therefore, produced a single-span level variation at channel W of 2 dB and a single-span level variation at channel 2 of 0.8 dB. If this effect is expanded to include 20 amplifiers in cascade, the total system variation at channel 2 and W becomes

Channel W 20 amps x 2 dB   = 40 dB

Channel 2 20 amps x 0.8 dB = 16 dB

It should be obvious that no system can withstand these extremes variations. In fact, if system levels exceed the overload level by only about 5 dB, picture quality deteriorates from excellent to totally unacceptable. At the other extreme, an uncompensated reduction in signal level of approximately 15 dB will result in unacceptable picture quality due to noise. (The amplifier operating window is adjusted to center the operating range, which allows relatively equal safety margins for noise and overload.) Consequently, a means of compensating for cable loss changes with temperature is an essential requirement of any cable system.

Changes in cable attenuation with temperature would be relatively simple to compensate if the change was linear with frequency; that is, if the same dB change occurred at all frequencies. This, however, is not the case. As the temperature changes, the same percent of change occurs at all frequencies; but, because cable loss with frequency is nonlinear, so is the resulting temperature-related cable loss change. In other words, cable slope or tilt changes with temperature. An example may help to clarify this point.

Refer to the previous calculations for derivation of numbers. A 20 dB, 300 MHz length of cable at 68°F exhibits a loss at channel 2 of 8.65 dB. If the temperature increases by 50°F, the cable loss at all frequencies will increase by 5%. Thus, 5% times 20 dB equals 1 dB and 5% times 8.65 dB equals 0.43 dB. Therefore, the level at 300 MHz has changed by 1 dB while the level at channel 2 has only changed by 0.43 dB. Consequently, the cable tilt or slope at 118°F is different than the tilt or slope at 68°F.

The cable temperature characteristic dictates that a temperature compensating amplifier must not only compensate for maximum gain change, but must also compensate for changes in cable tilt. This requirement, coupled with bandwidth and the temperature range over which the compensation must accurately track, represents one of the most critical design considerations necessary to ensure year-round system performance. The absence of this capability in a trunk system results in seasonal balance adjustments.

Some amplifiers have the inherent capability of both open-loop and closed-loop temperature compensation. The open-loop temperature compensation consists of slope and gain temperature tracking networks, which can compensate for characteristic changes in a 22 dB, 300 MHz span of cable over the temperature range of −40°F to +140°F.

Amplifiers which do not have this capability require some form of temperature compensating device.

12 Coaxial Cable Transmission Line Types

The primary component of any cable television system is the coaxial transmission line. The other components of the cable television system are used to get signals into or out of the cable, or overcome some limitations of the cable.

Coaxial cable is the most frequently used medium for signal transportation in existing cable systems. It is through the coaxial cable that signal energy is transported from an origination point to the input of the subscriber's TV set. If there were no loss in coaxial cable, signal energy could be transported infinite distances without the need of amplifiers. In reality, however, a particular cable at a particular temperature has an attenuation characteristic that is directly related to frequency. Furthermore, as the temperature changes, the frequency attenuation characteristic of cable also changes. It is, therefore, the frequency temperature attenuation characteristic of cable that dictates the need for equalized amplifiers and initial system balance and alignment. To fully understand these requirements, it is necessary to examine the frequency attenuation characteristics of cable and the temperature related changes that occur.

12.1 Basic Principles

The purpose of a transmission line is to carry electrical energy from one point to another with minimal loss. At low frequencies, almost any arrangement of conductors of the proper size will work. As the frequency becomes higher, it becomes harder to transfer energy through the cable. In most cable systems, the coaxial cable is divided into several functions, which are trunk lines, feeder lines, distribution lines, and subscriber drop lines. The cable used for these various functions may be different in size.

12.2 Center Conductor

The center conductor is a vital part of the coaxial cable. Most of the attenuation of the coaxial cable is determined by the center conductor. The larger the center conductor, the less attenuation that will be created.

There are four basic types of center conductors commonly used in cable television systems: (1) solid copper, (2) copper-clad steel, (3) copper-clad aluminum, (4) fiber optical. Fiber optical is considered to be the best conductor of these various types, solid copper the second best. Even though fiber optical cable is the best conductor, more systems use solid copper or copper-clad because fiber optical cable is much more expensive than the other.

Direct current flows through all parts of a conductor equally, however, alternating current is different. Alternating current tends to move toward the outer edge of the conductor, which is known as "skin effect." The higher the frequency, the higher the skin effect.

On a cold day, the copper used in the length of cable between active (amplifiers) and passive (directional couplers) devices may shrink more than the aluminum shield, resulting in the center conductor being pulled out of the connector. When this happens, a break in the line is created, and all signal and voltage is lost. This can be a real problem in cold areas.

The problem is not as severe with copper-clad aluminum center conductors. Inner and outer conductors tend to shrink at the same rate. Many connectors currently available have seizer screws which help hold the center conductor in place.

Fiber optical cable has no shrinkage at all because it is made of glass.

12.3 Strip Braids

In the early days of cable, in an effort to solve the temperature problem, strip braid cables were developed. This cable was made like ordinary braided shields, except that the shield was braided from relatively wide but thin copper strips, which decreased both radiation and attenuation. Strip braid cables were used for CATV systems until solid aluminum and copper shields were developed. Later the industry turned to an outer conductor that provided 100% shielding, which is now considered a standard.

12.4 Aluminum Shield

Cable with a solid aluminum shield is the cable most widely used in CATV systems. Solid aluminum provides 100% shielding. There are two types of solid aluminum sheathed cable; seamless and seamed.

12.5 Drop Cable Shield

Solid aluminum shielding for drop cable is seldom used. Drop cables are not as critical as trunk line cables and solid aluminum is very hard to work with in drop applications.

12.6 Thick Aluminum Tape

Instead of one or two thin layers of aluminum foil, some cables use a relatively thick (8 mm) aluminum tape. This tape is thick enough to provide 100% shielding and strong enough so as not to require a layer of braid. The thick aluminum tape is bonded to the cable jacket for moisture protection. The only disadvantage of this type of cable is that it is costly and not as flexible or easy to work with as other drop cables.

12.7 Braided Shields

There are numerous types of braided shield. You can get anywhere from 30% braid coverage to 97% braid coverage, in either aluminum, copper, or copper-clad aluminum. In general, the higher the braid coverage, the more you pay for the cable. Coverage of 50% or less is recommended only for systems without local signals and minimal interference sources. In large metropolitan areas, double layers of braid, or very high braid coverage, is required to eliminate problems of direct pick-up and radiation.

12.8 Dielectrics

In order to manufacture a coaxial cable, it is necessary to use some material between the center conductor and shield. From an electrical standpoint, the ideal material would be air or some inert gas. Unfortunately, this does not provide a means of holding the center conductor in the center of the shield. The earliest CATV coaxial cable was constructed with a dielectric solid tube, copper braid, and it still did not provide the strength necessary to avoid flattening. That function was accomplished by the solid dielectric. Later, when foamed polyethylene dielectric came into use, the cable was more subject to damage because it lacked the strength to resist being dented or flattened.

When aluminum sheath coaxial cable came into use, foamed polyethylene was used as the dielectric. The aluminum sheath provided considerable resistance to denting and flattening, although it could still be damaged by careless handling.

However, the dielectric was never intended to provided physical strength. Its major purpose was to keep the center conductor in the center of the shield. In the late 1960s, aluminum sheath coaxial cable, which used foam polystyrene dielectric, came into use. The electrical characteristics were better than foamed polyethylene, but the mechanical characteristics required careful handling. The aluminum sheath was more easily dented, kinked, and flattened than cable which had a foamed polyethylene dielectric.

The dielectric material originally used in drop cables was solid polyethylene, but foamed polyethylene soon came into use. Because the drop cable is small in diameter and is always jacketed, its resistance to denting and flattening is much greater than the larger cable. Also, drop cable uses a braid or foil shield which does not stay dented or flattened as solid aluminum does. However, the dielectric plays a significant part in the cables' resistance to denting and flattening. The drop is usually supported on each end by a device which grips the cable and applies an inward force. The dielectric helps to resist the tendency of the cable to be foamed at these points. For these and other reasons it has not been practical to use foamed polystyrene or air dielectric for drop cables.

Polyvinyl chloride (PVC) has the advantages of being flame retardant and more flexible than polyethylene, but it is contaminating. This means that the plasticizer used in polyvinyl chloride is relatively volatile. Over the years, some of the plasticizer is diffused into the air while some migrates to the cable dielectric. Therefore, the jacket becomes brittle and, in some cables, attenuation may increase. Cable covered with polyvinyl chloride ought to be replaced within five or ten years.

As its name suggests non-contaminating polyvinyl chloride uses a plasticizer that is not volatile, and therefore, lasts indefinitely. Non-contaminating polyvinyl chloride can also be buried, but it cannot operate in temperatures as low as polyethylene.

Technicians can become confused about the different kinds of jackets. For one thing, it is hard to tell them apart. For another, they have lots of different names and nicknames. Table 12.1 shows the characteristics of the different types of jackets and gives their most accepted nicknames and abbreviations. As you can see, both types of polyvinyl chloride are generally called "vinyl" and this often leads to misunderstandings. The safest thing is to refer to each jacket by its full name or abbreviation. The nicknames cause trouble.

The correct abbreviations are:

Polyethylene	PE
Polyvinyl Chloride	PVC
Non-Containing Polyvinyl Chloride	NCV

Table 12.1
Properties of Jacket Materials

Characteristic	Jacket Material			
	Polyethylene (PE) (Poly)	Vinyl (PVC) Type I	Vinyl (PVC) -50°C	Non-Contaminating Vinyl (NCV) Type IIa
Flame Retardant	No	Yes	Yes	Yes
Weathering Resistance	Excellent	Good	Good	Good
Crack Resistant	Yes	No	No	No
Abrasion Resistance	Excellent	Good	Good	Good
Contamination	Non-Contaminating	Contaminating	Contaminating	Non-Contaminating
Flexibility: Room Temp.	Fair	Good	Good	Good
Low Temp.	Poor	Fair	Good	Good
Direct Burial	Recommended	No	No	Yes

12.9 Moisture

Moisture poses a direct threat to CATV cable. Moisture in a cable can cause a tremendous increase in cable loss. Thus, customers who once received excellent picture quality complain of snow when the cable is wet.

This was a severe problem in the early days of CATV and technicians found this difficult to understand because they made their connections carefully, used plenty of waterproofing, and the cable was invariably jacketed.

Jacketed cable looks moisture-proof, but it is not. Water cannot go through PE, PVC, or NCV, but water vapor can.

In early CATV cables, using braided shields, water vapor passed through the jacket and condensed inside the cable. In rainy weather, water accumulated between the jacket and the shield, and then it ran down the cable to the bottom of its arc, where the water accumulated.

To verify this phenomena, one cable manufacturer tested polyethylene jacketed braided cable. The cable was sprayed thoroughly every day for 60 days. At the end of that time, results showed that attenuation had increased by 0.43 dB per hundred feet at channel 13.

Moisture resistance was one of the main reasons that the industry turned to aluminum sheathed cable. First, aluminum sheathed cable is completely waterproof. Aluminum will not allow either water or water vapor to pass. Second, if water does enter through pinholes or connectors, it still cannot travel through the cable because the aluminum is compressed around the dielectric. Moisture is even more serious a problem with underground cables. To keep moisture out, underground cables are sometimes made with a self-sealing flooding compound between jacket and shield. This compound seals any pinholes that may develop in the jacket, which is generally made of extra high molecular weight polyethylene. Sealing compound can be used on drop cables as well as trunk and distribution cables.

Aluminum sheathed cables are so rugged that no jacket is required in many CATV applications, but sometimes even aluminum sheathed cable must be covered. In shore and industrial areas, jackets are used to protect the aluminum from salt, air and other pollutants. Underground systems require jacket cables, and jackets are also recommended in areas with squirrels or other rodents. Rodents do not eat aluminum, but they love to sharpen their teeth on it. This can cause extensive damage.

12.10 Why 75 Ohms?

Use of these instruments, which are calibrated at 50 ohms, in 75 ohm systems will usually require an impedance matching device. Some instruments, such as signal generators, may be used directly in 75 ohm systems by calculating corrections to account for the impedance mismatch. Other instruments may be matched to 75 ohm systems by use of resistive attenuator pads designed to effect the required impedance transformation. Matching pads of this type, designed to make the 50 to 75 ohm match with minimum insertion loss, will exhibit very flat response over very wide frequency ranges due to their purely resistive nature. These matching pads will have a minimum insertion loss of 5.7 dB, which is the theoretical minimum insertion loss for a 1.5 ratio of impedances.

13   Powering the Cable TV System

In most cases, a cable television system covers a wide area, and many problems can arise in supplying operating power to the various components within the system. The amplifiers of a cable system are often spaced several thousand feet from each other and often cover an entire community. Each amplifier must be supplied with a specific operating power.

Many older cable systems used vacuum-tube amplifiers that required a power connection line at each amplifier location. This was an expensive arrangement which required most systems to be designed with a much greater than optimum spacing between amplifiers. The invention of amplifiers with transistors required much less operating power. Thus, it became practical to transmit the operating power along the same coaxial cable that carried the signal.

Transistors require a dc voltage, which requires the ac voltage to be converted to dc. It would seem logical to transmit dc voltage whereby conversion would not be necessary. However, this is not practical because of the corrosion which would result from electrolysis. A cable television system uses several different types of metals, such as copper and aluminum. When two different metals are connected in the presence of moisture and other atmospheric impurities, the connection tends to react like a small battery, thereby creating electrolysis. When a direct current passes through a series of such connections, it will aid some of the "batteries" and oppose others. Some connections will be worse than others. Most cable systems today use ac power transmitted along the coaxial cable to avoid such problems.

13.1 Coupling AC Power to the Coaxial Cable

There are three main considerations in providing power to a coaxial cable that also carries television and data signals:

1. The signal in the cable must not be shorted out through the power supply.

2. Noise and interference present in the power line must not be coupled into the cable system.

3. The configuration of the coupling device must be virtually transparent to loss and reflection.

The device used to couple power into the cable is called a *power inserter*. This device is constructed in much the same way as a directional coupler and signal splitter. It contains filters that allow the power to flow into the cable, but will block the signal from flowing into the power supply.

The coupler conductors are arranged in such a manner as to make it look like a section of 75 ohm cable to the signal. The electrical performance of a power inserter is described in terms of isolation between the cable and power source in decibels, i.e., return loss at its input and output.

The resistance values in Table 13.1 are typical of the common aluminum-sheathed coaxial cable used in CATV. These values of dc or low-frequency ac are given for resistance at a normal temperature (68° F).

These resistance values are for a typical aluminum-sheath, copper center conductor, foam dielectric cable. Center conductor resistance for copper-clad aluminum center conductors is somewhat higher, and, consequently, loop resistance is also higher. Cables which have an air or foam dielectric with a

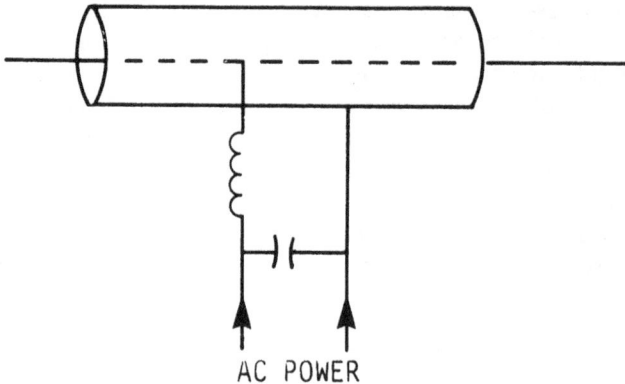

Figure 13.1 AC Power

Table 13.1

Nominal Resistance

(ohms per 1000 feet)

CABLE TYPE	OUTER CONDUCTOR	INNER CONDUCTOR	LOOP
0.412 in.	0.43 ohm	1.59 ohm	2.03 ohm
0.500 (1/2) in.	0.35 ohm	1.08 ohm	1.43 ohm
0.750 (3/4) in.	0.17 ohm	0.45 ohm	0.62 ohm

higher air content than the commonly used foamed polyethylene dielectric generally have a center conductor with a greater diameter for the same size outer conductor. Consequently, these have a lower center conductor resistance and a lower loop resistance.

In practice, the power source stays constant and a drop in voltage at the amplifier occurs due to the IR voltage reduction in the cable loop resistance.

Most CATV amplifiers draw nearly a constant current over their normal range of ac operating voltages. This is one of the more undesirable characteristics of most CATV amplifiers, and represents a departure from Ohm's law, which states that current flow is directly proportional to applied voltage ($I - E/R$). This constant current behavior is because of the operation of the series regulator circuitry commonly used by CATV amplifiers.

The amplifier works well with only 59.38 volts instead of 60 volts. In fact, it could continue to work properly with as little as 44 volts. This wide range of ac-voltage input is handled by the regulating circuitry inside the amplifier. There is minimum ac voltage at which the amplifier will operate properly. More practical examples illustrate that this minimum operating voltage is a serious limiting factor in the design and operation of CATV power systems.

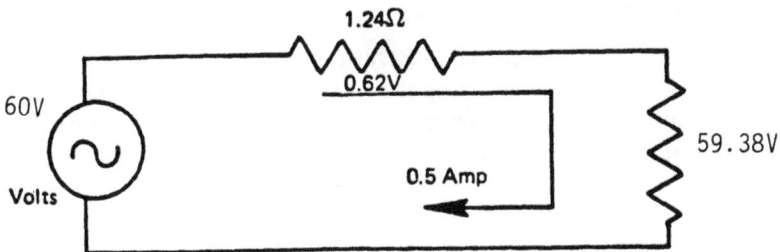

Figure 13.2 Resistance

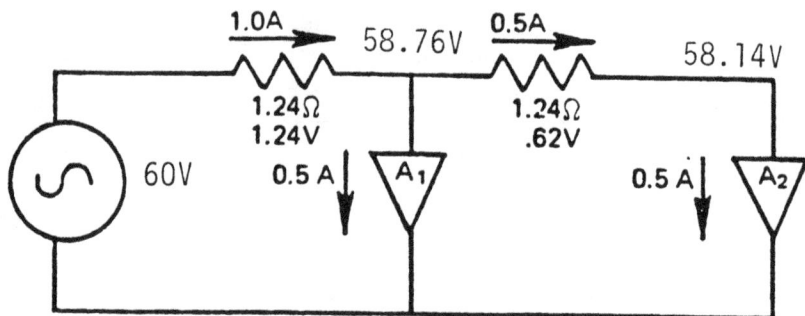

Figure 13.3 Voltage Drop

Note that in Fig. 13.3 the first cable section is carrying 1.0 amp, which is the current for both amplifiers. The second section carries only the current for the second amplifier, 0.5 amp. The voltage reduction in each cable section has been calculated by Ohm's law ($E = IR$); thus, the voltage drop is the current times the resistance. With 60 volts available at the power feed point, the first amplifier has 58.76 volts, and the second has 58.14 volts.

The technique for power calculation is simple and straightforward:

1. Prepare a detailed power flow schematic diagram showing all the amplifiers and cables which carry power.

2. Enter the cable section loop resistances and the amount of amplifier current drawn.

3. Calculate the current flow in each cable section, summing currents of all the amplifiers fed through the cable for each cable section. Start from the farthest amplifiers and work back toward the power feed point.

4. Calculate the voltage reduction in each cable section and mark it on the diagram.

5. Starting at the power feed point, calculate the voltage at the end of each cable section by subtracting the voltage reduction from the voltage at the input. Note that power often flows in the opposite direction of signal flow.

6. Verify that each amplifier has an adequate operating voltage. Minimum operating voltage varies among manufacturers and amplifier types. An amplifier designed for nominal 60 volt operation often performs adequately on a minimum of 44 volts.

7. Make sure that the current draw does not exceed the current rating of the power inserter. The greater the current draw, the greater the possibility of hum problems.

If any amplifiers seem starved for voltage, some rearrangement is required. Low voltages will cause hum bars in the television picture and other operating problems. Certain amplifiers may have to be transferred to another power feed point, taking care to avoid overloading the power distribution facilities. Some power reserve must be allowed so that additional line extenders, active taps, amplified splitters, *et cetera*, can be accommodated in the future without overloading the cable power system.

13.2 Diplexing

The capacitor is chosen (usually 0.01μF) to present a low impedance to RF, but to present a high impedance to 60 Hz power. A 0.01μF capacitor has a reactance of only 3.2 ohms at 5 MHz, the lowest frequency that is ever expected to be handled in the RF section of a CATV system. At 50 MHz, which is a more likely lower frequency limit for CATV RF, the reactance of a 0.01μF capacitor is only 0.32 ohms. The 0.01μF capacitor has a reactance of approximately 270 thousand ohms at 60 Hz power line frequency. An inductance of 1 millihenry has a reactance of approximately 30 thousand ohms at 5 MHz, but only 0.37 ohms at 60 Hz power line frequency. The reactance of this inductance rises to 300 thousand ohms at the more common RF of 50 MHz.

The diplexing arrangement (Fig. 13.4) is shown with typical values for the capacitor and inductor, with the reactance at 50 MHz and 60 Hz. The mixing and separation of RF and power can be more easily understood by following the behavior of the diplexing network at power frequency and RF. Power will follow the path of least resistance from the power to the RF/power terminals through the 0.37 ohm reactance of the inductor (RF choke).

Power is kept from the RF terminal by the 270 thousand ohm reactance of the capacitor. RF also follows the path of least resistance, passing easily through the 0.32 ohm of the capacitor, but is blocked from the power terminal by the 300 thousand ohms represented by the inductance. The diplexer effectively separates or combines RF and power.

Actual diplexer circuitry may differ slightly from the basic arrangement discussed here. The capacitor can be omitted, allowing RF and power to appear on both ports. Two inductors may be used with small capacitors to ground, so that no RF can possibly reach the power terminal. Such a combination may be used to permit separation of power from input and output connectors.

13.3 Amplifier Power Supplies

The purpose of cable power is to provide the energy for the RF amplifier. The amplifier power supply must change the ac cable power to dc of the proper polarity, and provide adequate filtering and regulation. Some amplifiers require dc voltages of both polarities. Most require voltages of only one polarity, either positive or negative with respect to ground, depending on the details of circuit design and the types of transistors used. The dc power supply for the amplifier must be well regulated, because amplifier gail will change if the dc supply voltage changes. The dc power supply must also be well filtered; thus, the dc is pure.

13.4 Full-Wave Power

Transformer type supplies are now commonly used, at least in larger CATV amplifiers. They permit simple, full-wave receiver circuitry, and the ability to use tapped primary windings permits more efficient supply of cable power. Excessive rectifier output voltage is dissipated in the regulator as heat (except in switching type regulators). Availability of primary transformer taps permit the technician to adjust the rectifier output voltage to an optimum value and also allows the regulator a reasonable operating range without excessive power wastage through heat dissipation. Transformers also provide for operation over a wide range of cable voltages. Some systems operate with nominal 60 volt ac power instead of the usual 30 volt ac power. The use of transformers permits an easy transition from 60 to 30 volt operation, simply by alternating transformer taps.

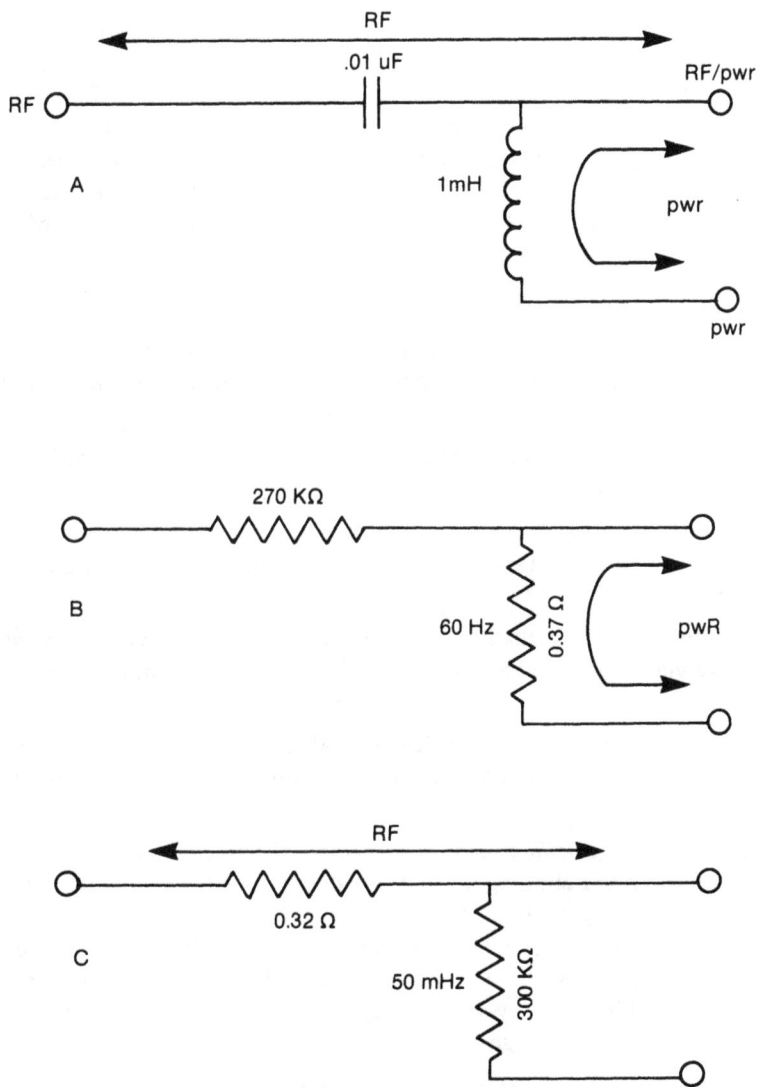

Figure 13.4 RF Power

Direct current would be the most efficient way to carry power on the cable. Power would flow smoothly and continuously. Amplifiers would not need rectifiers and filters, but would still require good voltage regulation. It would be easy to provide stand-by power in case of general power failure — a simple storage battery would be sufficient. However, dc power is not used because of the great danger of corrosion at connectors and fittings due to electrolytic effects. Direct current in the presence of moisture can act as an electroplating current and cause serious damage to the CATV operating system.

13.5 Transients and Power Surges

CATV systems must be protected from the damaging effects of voltage transients and surges in the cable. Abnormal operating voltages fall into three basic categories:

1. *Short duration*, usually 2 to 20 times normal voltage. Generally caused by lightning strikes or lightning-induced power surges.

2. *Medium duration*, usually 2 to 4 times normal voltage. Generally caused by turn-on surges of commercial power or power supplies of the CATV system.

3. *Long duration*, usually 2 to 20 times normal voltage. Generally caused by accidental contact between cable system and commercial power lines.

There are many surge protection devices that can be used in various combinations to achieve a high degree of surge immunity. Some of the more useful devices are:

1. *Miniature gas-filled surge suppressor*. These devices offer low capacitance bipolar clipping for voltages over a specific value. Lower breakdown voltages are not possible because of the characteristics of Paschen discharge in low pressure gasses. Gas-filled surge suppressors have an intermediate response time and can handle larger current surges of low duty cycle, but they do not stand up well under continuous current discharge over 1 A.

2. *Zener diodes*. These devices have intermediate capacitance values (30 pF for 1 W devices) and relatively fast switching time. They have lower surge current capacity because of their power dissipation limitation. Breakdown voltages range from 1 to 200 V and zener diodes are frequently used to trigger other devices which can handle larger surge currents.

3. *Solid state switches*. These devices usually rely on external circuitry (resistors and zener diodes) to provide accurately controlled turn-on. They have moderate power dissipation capacity and are widely used for power supply protection.

4. *Variable resistance devices*. (a) *Varistor*. These are metal-oxide devices that have a voltage variable resistance. They have high capacitance, fair peak capability, and breakdown voltages range from 30 to 400 V. Limiting characteristics are fair. (b) *Thyristor, Thyrector, Thyrite*. These are mainly silicon carbide devices. They have high capacitance and good surge current capacity, but poorer limiting characteristics than varistors.

5. *Mechanical switches*. During a power surge, these devices disconnect the applied power. They are slow acting and can be programmed to reconnect power to the system following a prescribed waiting time after the return of primary power.

Because surges can be severe, and can be introduced at any location in the system, it is desirable to place protection devices at scattered locations throughout the system. These protective devices minimize or attenuate the surge injected into the system and protect the electronic equipment from any residual surges that will inevitably intrude at a later time.

Some of the protection points and associated methods are:

1. At the *power input*. Filtering and a high energy device to absorb transients. Attention to grounding to minimize ground coupled transients.

2. At the *power supply output*, where the transformer transients are introduced. Use time-delay relays or energy absorbing devices to limit the output voltage.

3. At the *amplifier input and output*. Gas diodes attenuate transients that otherwise could be capacitively coupled to the RF transistors.

4. At the *secondary side of the transformer of the dc power supply* (which is highly vulnerable because of its low frequency, direct connection path to the transformer). This is the only protection point that prevents sheath currents from affecting the amplifier station.

13.6 Power Circuit Protection

Cable power circuits are always protected by a series of fuses or circuit breakers. These protective devices must be used in a logical manner in order for them to do their job properly. Generally, fuse or circuit breaker values increase along the power circuitry from the most distant equipment back toward the power source. This gradual increase prevents minor problems on distribution lines from blowing fuses at critical points in the system and, thus, disrupting service to large portions of the system when it should have affected only a small part of it.

Consider a simplified schematic diagram of a portion of a cable power layout. The main fuse F1 must be capable of carrying all current drawn from the power feed. This fuse is backed up by a power circuit breaker or a fuse on the primary side of the cable power transformer. The primary circuit breaker protects the utility company's system against any short circuit in the cable system or the power feed transformer. Fuses F2 and F3 protect individual main line amplifiers. If any of these amplifiers develop a short, the fuse within the amplifier will blow. Therefore, these fuses must be lower in value than F1. Otherwise, a short in the farthest trunk amplifier would blow F1, the main fuse, before it blows its own amplifier fuse. Everything connected to that power feed would then be out of service. If only F2 blew, then at least part of the system could still be working.

F4 is also considered to be a main fuse because it protects an entire distribution line with several line extenders. The fuses in each line extender (F5) must be lower in value than F4. Otherwise, a short in the farthest line extender would blow the main distribution line fuse and then blow the rest of the fuses along the line.

We see that the size of the fuses must be carefully controlled and designed with regard to the nature of the device or system being protected. Fuses must be graded in size in order to provide a logical chain of protection to the system.

Tracing dc short circuits in a cable power system is difficult and time consuming. If the fuse configuration has been logically designed, locating the position of the blown fuse in the system can be done easily and the trace can proceed efficiently. If the main fuse F1 has blown, it is unlikely that the short is in an individual amplifier or on one of the distribution legs because these are all protected with fuses smaller than the main F1. Similarly, if one of the distribution line fuses (F4) has blown, it is unlikely to be one of the line extenders because these are individually protected by fuses smaller than F4. If the fault had been in one of the line extenders, its own fuse would have blown before the major fuse preceding it. This emphasizes the importance of proper planning of fuses, and the importance of replacing fuses with the proper values in a timely fashion.

A helpful device for locating short circuits is a small light bulb attached to a dummy fuse. A 60 volt, 10 watt (or even lower wattage) light bulb can be wired into a fuse plug so that it takes the place of the fuse when plugged in. The resistance of the lamp limits the short circuit current to a safe value and the lamp remains lit until the short circuit is cleared.

Indicating fuse holders that use this same principle are also available. A small lamp of suitable voltage rating is wired across the fuse. When the fuse blows, the lamp lights up because the short completes its circuit. The lamp goes out when the short circuit is cleared. It is not necessary to replace every fuse holder with this indicator type, but a special fuse holder can be devised with a small lamp that provides this test feature when desired.

13.7 AC Power Supplies

Alternating current power supplies for CATV systems are preferably of the regulating transformer type. This transformer furnishes the power which is supplied to the cable for system operation. Units have high reliability and fully protected circuitry, combined with overload and transient protection, which will operate into any load, up to short circuit, without damage. In some cases, plug-in modular construction is used for each transformer in field service and installation. Unit housings are weather-proofed steel with provisions for direct mounting on utility poles. Low profile pedestal amount units are available for underground systems.

13.8 Summary

It can be seen that powering of CATV amplifiers is a complex procedure. If the current through F1 is 4.5 amperes and a 5.0 ampere fuse is used, another 0.5 ampere of current will cause F1 to blow. If F3 normally has 1.0 ampere flowing through it and a 2.0 ampere fuse is used, it will take another 1.0 ampere to blow F3. A component failure in the bridger fed by F3 could cause it to draw an additional 0.75 ampere causing 1.75 ampere to flow through F3; but the 2.0 ampere fuse would not blow. However, an additional 0.75 ampere would then flow through F1 and the total of 5.25 amperes would cause the 5.0 ampere fuse in that location to blow, which would turn off all the amplifiers receiving power from this ac location. Obviously, this is a most undesirable event. The selection of fuse values is, thus, very important, but not a simple process. In the above example, if F3 were 1.25 ampere, it would blow at the moment the failure occurred in the amplifier.

F1 should be 6 amperes so as not to blow. When F3 blows, the 1.0 ampere which the amplifier normally used would stop flowing. With less current flowing, the voltage drop through some portions of the cable would be less, resulting in a voltage increase to the amplifiers powered by the ac power supply.

With accurate powering maps or data sheets, it is often possible to determine which amplifiers are not working simply by knowing which subscribers have called to report a signal outtage. Then it may even be possible to predict which fuse has blown. There are too many possible power failure symptoms to discuss here. Remember that a blown fuse in a line extender will only affect the subscribers who receive their signal from that extender, or any extenders which follow in the same feeder line. A blown feeder line fuse will only affect the signal beyond the first line extender, but a blown bridger fuse will stop the signal into all the feeder lines from that bridger. None of these fuses will affect the signal on the trunk. A blown fuse in a trunk amplifier will, of course, stop the signal on the trunk at that point. A blown fuse in the ac power supply (F1) will stop the signal at the trunk amplifier farthest up the system (toward the head-end) which is powered from that location. A good understanding of powering techniques and familiarity with the powering of your particular system will make it much easier and faster to locate blown fuses and other powering problems.

14 Passive Devices

A *passive device* is one which exhibits no transistance. It has no gain or control and does not require any input other than a signal to perform its function. Passive devices do not provide rectification, amplification, or switching, but react to voltage and current.

14.1 Taps

Taps are passive devices normally placed in the distribution or feeder line to extract signal from the feeder line for the subscriber television. A tap may have anywhere from two to eight ports that the subscriber drop can hook up to. The insertion loss of a tap varies with the tap value. Normally, as the tap value increases, the insertion loss will decrease. Taps are usually designed to have a frequency response of ±.25 dB across the total bandwidth and the capability to pass power with minimum attenuation. Isolation on taps should be >26 dB between the port and tap output. *Directional couplers* and *power inserters* are also passive devices.

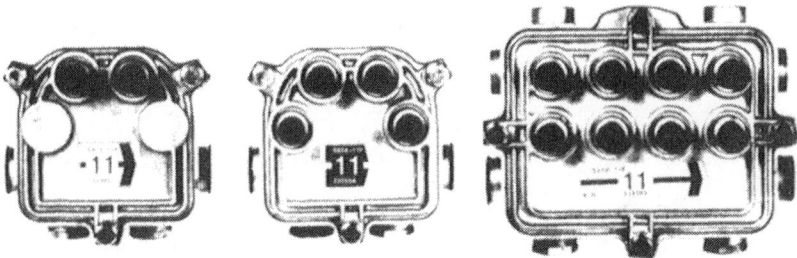

Figure 14.1 Taps
Source: Scientific Atlanta

14.2 Splitters

If two 75 ohm coaxial transmission lines, such as those used in CATV systems, are connected in parallel, the resulting load will be 37.5 ohms.

Even though the power will divide equally between the two branches, and the RF voltage will be the same in each branch, a two-to-one impedance mismatch exists at the junction. One-half the power will be reflected back toward the input — and absorbed by the 75 ohm source impedance. It is, therefore, important that the conventional 75 ohm impedance used in CATV transmission systems be properly matched.

When it is necessary to divide a CATV transmission line, special devices are used to maintain the impedance match. These devices are commonly referred to as *line splitters*, or simply *splitters*, and are so designed as to present a 75 ohm characteristic impedance to all branches of the line. A two-way line splitter is constructed with tapped transformers in a circuit configuration designed to equally divide the power between the two branches while maintaining the 75 ohm characteristic impedance in all branches.

Three-way and four-way line splitters are actually combinations of two-way line splitters. For example, a three-way line splitter consists of a two-way line splitter with one of the branches being split again. Therefore, the branches of a three-way line splitter will have outputs that are 6 dB down on two of the branches and 3 dB down on one of the branches. A four-way splitter will have both branches split, and thereby all four output branches will be down 6 dB. Of course, these are the ideal or theoretical values, and in actual practice the losses will be somewhat greater depending upon the Q factor of the coils and efficiency of the transformer circuitry. Actual values are usually about 3.25 to 3.50 dB loss in each two-way split —a four-way splitter having outputs that are 6 ½ to 7 dB down from the input level.

15 Concepts of Distribution Systems

A cable television system provides the *link* between the head-end and the component sub-systems. The primary functions of this link are to carry the various television signals from the point of reception, or origination, to convenient points for distribution to subscriber locations and connection of individual subscribers. The complete system is conveniently divided into three sub-systems:

1. The trunk or transmission systems

2. The feeder or distribution system

3. The subscriber drop, or connection

15.1 The Trunk System

The *trunk system* extends from the origination point, or head-end site, to the various distribution centers where signals are furnished to the distribution system by isolation devices. The trunk system is also referred to as the *transmission system*.

Because trunk system performance affects the entire system, no subscriber taps or direct connections are made to the trunk cable. Every effort is made to preserve the trunk system signal quality and reliability as the system's maximum size is determined by the length of trunk cable which can be operated without objectionable video or audio degradation.

The trunk system consists of low loss coaxial cable, repeater amplifiers with automatic control features which provide correction for the effects of temperature on the cable's attenuation, and bridging amplifiers for isolated signal feed to the distribution system. The trunk system may split into several directions from a central location for maximum area coverage with minimum signal degradation.

15.2 Hum Modulation

Another source of possible signal degradation is *hum modulation*. Amplification system hum could be introduced into the system by faulty rectifiers (heater-to-cathode leakage), or by ineffective dc power filtering. The source, generally in the amplifiers, is easily located and corrected. In contemporary CATV systems, amplifiers are cable powered with low ac voltage and are completely transistorized. The sources of hum modulation in a contemporary system include:

1. *Low ac voltage on the cable powering system.* This results in loss of the zener regulation employed in the amplifier's dc supply and a consequent increase in the ripple component appearing in the dc applied to the various transistors.

2. *Current saturation of the ac power bypass coils used in amplifiers and passive devices.* Iron or ferrite core coils are used to prevent loss of RF signals into the ac power circuitry as well as to prevent the ac power from flowing through the RF circuitry. These coils fail to function when excess current is allowed to flow through them, causing hum modulation of the RF signals on the system.

3. *High resistance connections due to corrosion of connector contacts.* The effect of corroded connector contacts may result in their functioning as series diodes — with a consequent "diode modulation" of the ac cable power onto the RF signals.

4. *Induced ac from particularly strong magnetic fields.* Unshielded RF conductors or components in the vicinity of high-intensity ac power devices may introduce slight amounts of hum into the system.

The design of a CATV system involves determining the most efficient use of the selected cable and components to provide service to the area being considered. As the trunk system is not tapped for subscribers and only serves to provide transmission to the distribution points, it should be as short and direct as possible. The use of lower loss cable will not reduce the cost of the system, as the increased cost of the larger diameter cable, larger connectors and fittings, installation, *et cetera* will result in an increased cost per mile of the trunk system.

In order to accommodate the coverage of larger areas without degradation, it is sometimes necessary to use larger (and more expensive) cable, which will reduce the number of amplifiers in cascade to a manageable quantity. Normal practice is to use ½ foam dielectric aluminum-sheathed coaxial cable with a loss of approximately 1.3 dB per hundred feet at channel 13 for trunk lines extending up to about ten miles. Amplifier cascading limitations generally require ¾ cable for trunk lines longer than about ten miles. Three-quarter inch foam dielectric aluminum-sheathed coaxial cable has a nominal loss of 0.9 dB per hundred feet at channel 13 at 70°F. If even less cable loss is essential, there are larger foam dielectric cables available, and air dielectric cables may also be used to decrease the cable loss.

15.3 The Distribution System

The *distribution system* extends from the trunk system to the subscriber connection. The primary objective of the distribution system is to provide undergraded signals of adequate level to the subscriber's television (or FM) receiver. An acceptable subscriber level is 2 to 8 dBmV. Input levels lower than 0 dBmV may result in snow as a consequence of the receiver's noise figure, and input levels in excess of 10 dBmV may result in overload of the receiver's input circuitry, which produces cross-modulation or adjacent channel interference.

In order to provide the required signal level at the subscriber's connection point, the tap is selected to attenuate the signals to approximately 12 dBmV. These taps have either flat or tilted response characteristics so as to result in an approximately flat level of the various channel signals at the subscriber's receiver.

The highest practical undistorted output level should be used in the CATV distribution system. This will permit the greatest use of high value (low insertion loss) taps, and will reduce the requirement for additional amplification. By tradition, the amplifiers used on the distribution or feeder system are referred to as *line extender amplifiers*.

The distribution system is connected to the trunk transmission system through *bridging amplifiers*. These bridging amplifiers, or bridgers, are so designed as to provide considerable isolation to the trunk system.

The bridging amplifiers are fed from a directional coupler circuit, which is inserted in the trunk cable, of sufficient gain to provide the high levels required on the distribution system. Provision for up to four distribution outlets is included to accommodate the various distribution arms necessary to feed the area being served. The ac power fed through these arms is separately fused to isolate any possible trouble within the particular arm, which may have a defect or accidental short circuit.

15.4 Degradation of the Color Characteristics

Color degradation does not ordinarily occur on the CATV distribution system. *Differential phase distortion* involves a nonlinear change in the phase of the 3.58 MHz color sub-carrier signal as the television signal level changes over the chromatic scale from black to white. *Differential gain distortion* involves a similar nonlinear change in the amplitude of the 3.58 MHz color sub-carrier signal. *Envelope delay*, also known as group delay, is a nonlinear change in phase delay coupled with a linear change in frequency. These three factors are basically video limitations, and as such are relatively small and very difficult to measure on CATV distribution systems.

15.5 Evaluation and Performance Testing

Testing of the CATV distribution system involves the following factors:

1. Noise
2. Cross-Modulation
3. Hum Modulation
4. Power Supply Regulation
5. Frequency Response
6. Automatic Gain Control (AGC)
7. Subjective Signal Quality Evaluation

Noise tests are made to determine the amount of interference on the system generated by thermal agitation effects in the amplifiers. An ideal 75 ohm, 4 MHz television channel will have an irreducible level of noise equal to about −59 dBmV at 70°F. This noise will be increased by the noise figure of the first amplifier. If subsequent amplifiers have identical noise figures, the noise will then increase by 3 dB every time the number of cascaded amplifiers is doubled. This noise factor determines the minimum input signal level which may be allowed, depending upon the number of repeater amplifiers operated in cascade.

The objective of a noise test on a CATV system is to obtain a measurement which will indicate how far the level of the noise interference is below the operating signal levels of the various TV channels carried on the system. The measurement is actually made in terms of "carrier-plus-noise to noise" (C/N) rather than "signal-to-noise" (S/N) as it is sometimes called. Peak carrier values as measured on the system are used as a reference, requiring the use of a substitute unmodulated CW carrier in order to avoid the fluctuations resulting from the video modulation changes that occur on TV channels operating on the system. The signal levels are normally measured using an unmodulated CW carrier adjusted to the normal operating level of the system, and then repeated with the carrier removed from the system and the system input terminated with a 75 ohm impedance.

Table 15.1

Measured Ratio of CW Unmodulated Carrier

MEASURED RATIO (dB)	PICTURE QUALITY
over 42	EXCELLENT
40 to 42	GOOD TO EXCELLENT
38 to 40	FAIR TO GOOD
36 to 38	PASSABLE TO FAIR
34 to 36	OBJECTIONABLE TO PASSABLE
under 34	UNUSABLE

Cross-modulation testing measures the "cross talk" which interferes with the desired television channel. Initial indication is a non-synchronous horizontal blanking bar which creates the effect of a horizontally moving vertical bar similar to a windshield wiper on an automobile. This "wiper effect" is caused by a spurious horizontal synchronous pulse leaking into one channel from another, as a result of overload distortion of the repeater amplifier.

To establish a cross-modulation ratio requires expensive instrumentation and difficult measurement techniques. An indication of the system's amplifier overload distortion may be obtained to establish a "cross-modulation ratio" by providing synchronous modulated signals on all channels except the particular channel under test. A 15,750 Hz square wave is used for modulation of all channels except the channel under test, which is an unmodulated CW picture carrier. The test equipment is set up to measure the amount of the 15,750 Hz modulation that leaks into the unmodulated channel frequency from the other channels carried on the cable system.

Hum modulation is a form of interference which produces alternate dark and light horizontal bars across the TV screen. These bars may roll slowly up or down through the picture. One bar indicates a 60 Hz spurious signal and two indicates 120 Hz interference. Hum modulation of 60 Hz may be caused by inadequate dc power supply filtering of half-wave rectification circuits or from saturated bypass coils. Hum of 120 Hz is usually a result of low voltage (ac) furnished to the amplifier power supplies, causing them to cease regulation.

The usual test for this interference is to tune a connected signal level meter to a channel, detect and rectify the video information, and display the rectified output on a sensitive oscilloscope. Hum modulation of 2% of the peak video level is generally regarded as acceptable, with 1% or less being practically undectable.

Power supply regulation tests are made to determine the vulnerability of the system to ac voltage variation. Operating systems are often furnished ac power at a nominal 110 volts RMS, but the actual voltage may vary as much as ±20 volts because if existing power company load conditions, as well as for gain stability while the ac power source is varied between 90 and 130 volts RMS. The 115/60 volt regulated transformer supplying the cable system should accommodate this input variation.

Frequency response of CATV system must be maintained within reasonable limits to prevent gross differences among the levels of the various channels at the system extremities. Individual television channel response should be flat with about ±1 dB for undergraded reception, and ±3 dB flatness is usually desired across the entire bandwidth. Some operators express this in terms of $N/10 + 1$, where N is the number of cascaded trunk amplifiers.

A summation sweep-frequency test is necessary to achieve the desired level of response flatness. This involves the replacement of the regular TV channel source at the head-end with a swept-frequency RF generator and slight amplifier alignment correction for very small response variations, which are displayed at the amplifier locations by connection to a sweep receiver and oscilloscope. After careful calibration at the head-end, a reference trace is marked on the oscilloscope and subsequent displays are compared to this trace. Two problems occur in performing this test: one is a requirement to synchronize the oscilloscope's horizontal deflection with the frequency sweep of the generator source; and a second problem involves simultaneous use of a pilot carrier for AGC action and the swept frequency. If a trap or filter cannot eliminate the interference of the pilot carrier and swept frequencies, it may be necessary to perform this test manually. Synchronous horizontal sweep can usually be achieved if the sweep generator is linear in its frequency variation. Where a sinusoidal frequency change is used, it may be necessary to obtain line voltage for the oscilloscope's horizontal input —and this may be slightly out of phase.

A second method is to insert a calibrated signal generator into the system at the head-end, then read the levels at the test location at the far end of the system, and investigating the entire bandwidth at one or one-half MHz intervals. This is a cumbersome task requiring close coordination between the measurement point and the head-end. It permits the system response to be measured with the amplifiers operating normally.

A valid test of *automatic gain control* (AGC) action would require an extended period of time. To perform this test, a record would be made of the various channels' signal level variations over a period of several days, and another record would be made simultaneously of the temperature changes. It is essential that these records be accumulated over a period of time in which the temperature varies by at least 50°F. Such test environment conditions would indicate adequate control of the system levels through the effects of an average change in temperature from early morning to mid-afternoon. As many channels as possible should be monitored in order to record any response variation with AGC action, but at least one low-band and one high-band channel should be recorded to observe the control of amplifier tilt *versus* the temperature-induced change in attenuation and slope. The measurements must be unaffected by changes in measurement equipment line voltage in order to obtain valid data.

16 Theory of the Heterodyne Converter

16.1 General Theory

The *heterodyne converter* essentially consists of three major parts: the down-converter, IF amplifier, and up-converter. The down-converter, equivalent in function to the tuner in a TV set, converts the incoming signal frequency to the intermediate frequency (IF) range. In the IF amplifier, adjacent channel filtering is accomplished, the picture carrier level to sound carrier level is established, and the AGC circuit controls the gain of the IF amplifier to maintain a constant output level. The up-converter mixes the IF amplifier output to the desired VHF channel output. This is a very general description of the heterodyne converter. The detailed theory of heterodyne converter operation is given below.

16.2 Detailed Theory of Operation

A block diagram of a heterodyne receiver commonly used in CATV systems is shown in Fig. 16.1.

Down-converter section includes:
    RF pre-amplifier,
    local oscillator, and
    mixer;

IF amplifier section includes:
    video IF amplifier,
    AGC amplifier, and
    sound carrier IF amplifier;

Up-converter section includes:
    mixer,
    local oscillator, and
    linear RF power amplifier.

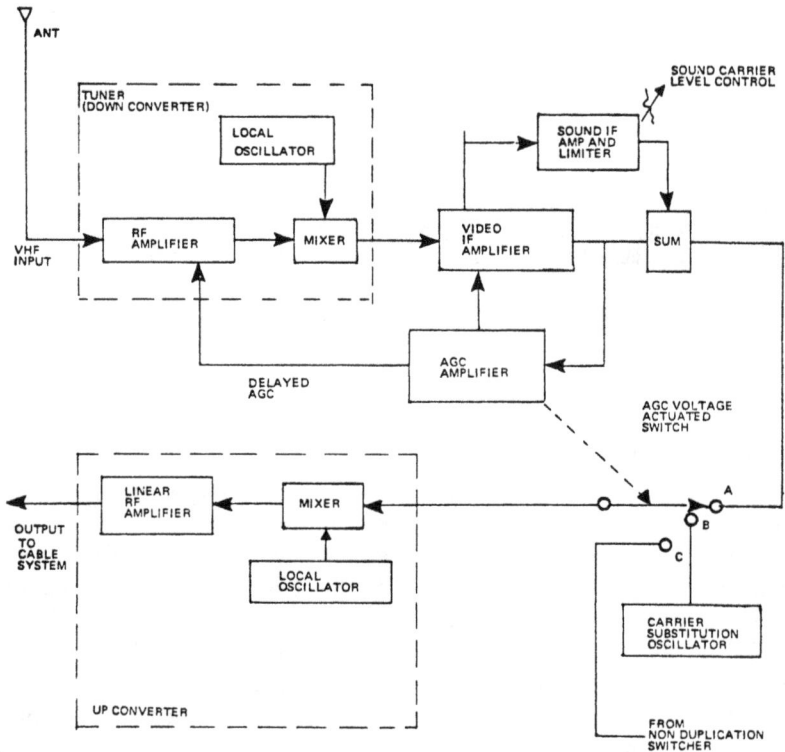

Figure 16.1 Detailed Block Diagram of Heterodyne Converter
Source: Scientific Atlanta

The incoming signal from the antenna is amplified in the RF pre-amplifier, and beats (or heterodynes) with the local oscillator frequency in the mixer, the output of which is now at IF frequency. Adjacent channel filtering is accomplished in the IF amplifier. The sound carrier at IF is separated from the television signal, amplified and limited, then summed with the output of the video IF amplifier. Prior to summation, the ratio of picture carrier level to sound carrier level is established by adjusting the sound carrier level. The output of the IF amplifier is held at a constant level by AGC action, which controls both the RF and IF amplifier gains. The RF amplifier is normally allowed to operate at maximum gain until the input signal level reaches a point at which the signal-to-noise ratio will not suffer by a reduction in gain. This is known as *delayed AGC*. The combined IF output (picture and sound carrier) is converted to the desired VHF channel by the output local oscillator in the output mixer. If on-channel conversion is required, the input local oscillator drives the output mixer and the output local oscillator is disabled. Differential gain and differential phase will usually occur, because mixer conversion is not related to modulation percentages.

The video carrier substitution oscillator is normally activated by the AGC voltage when the input signal level from the antenna drops below a predetermined level. This CW carrier is then converted up to the picture carrier frequency of the output channel. This feature provides two things. First, a carrier frequency is always on the cable system so that system levels can be checked if necessary. Second, if a composite AGC system is used on the cable system, the AGC system will operate normally even if the input signals received at the antenna go off the air.

16.3 Input Impedance

The input impedance to the tuner normally varies as a function of input signal level. This is because the first stage of the RF amplifier is usually in an AGC mode and its input characteristics vary as a function of AGC voltage. This input impedance match should be a minimum of 16 dB return loss over the expected range of input signal levels. This is important because a mismatch among the antenna, low loss down-lead, and heterodyne receiver will generate ghosts on the received channel.

16.4 Sensitivity and Maximum Level

The sensitivity and maximum level specifications define the dynamic range of the receiver. No specification on the sample data sheet lists the maximum level at which an adjacent channel can appear at the receiver input terminals without causing distortion in the desired channel. Nevertheless, the maximum adjacent channel level relative to the desired channel is an important performance specification for many applications. The wise technician should check with the equipment manufacturer if this specification is not listed.

16.5 Noise Figure

The noise figure is usually specified at maximum gain, which corresponds to minimum input signal. The noise figure is usually dependent only upon the tuner. The signal-to-noise ratio between input and output terminals of the heterodyne may not follow directly from the noise figure listed; indeed, it may be worse. This condition can prevail if the equipment has not been designed to account for the minimum signal levels throughout the receiver under all operating conditions. For example, if the output level control reduces the output level 10 dB from its maximum value, the noise figure of the up-converter mixer may be of such a value as to degrade the signal-to-noise ratio more than the listed specification of 6 dB.

16.6 AGC Sensitivity

The automatic gain control sensitivity specification may refer only to the picture carrier level, i.e., $-\frac{1}{2}$ dB change at output for 54 dB change at input. Be sure to determine the frequency response across the television channel as a function of input level change. Some systems can have extreme tilt in frequency response across a 6 MHz bandwidth over a 50 dB AGC range.

16.7 Output Level Range

Determine if output level range control is accomplished outside of the IF amplifier AGC loop. Some heterodyne receivers rely on changing the AGC threshold to adjust output level and, thereby, reduce the useable range of the AGC loop.

16.8 Adjacent Channel Filtering

Adjacent channel filtering is a key requirement of the head-end signal processor. One advantage of a cable system is that it can provide channels to its subscribers which cannot be received by the ordinary roof-top antenna. If these additional channels are not locally originated, they are probably received from a distant source. The channels received by the CATV antenna will typically occur adjacent to the local broadcasting channels. Antenna design provides directivity and some adjacent channel filtering, but the CATV receiver must provide most of the adjacent channel filtering.

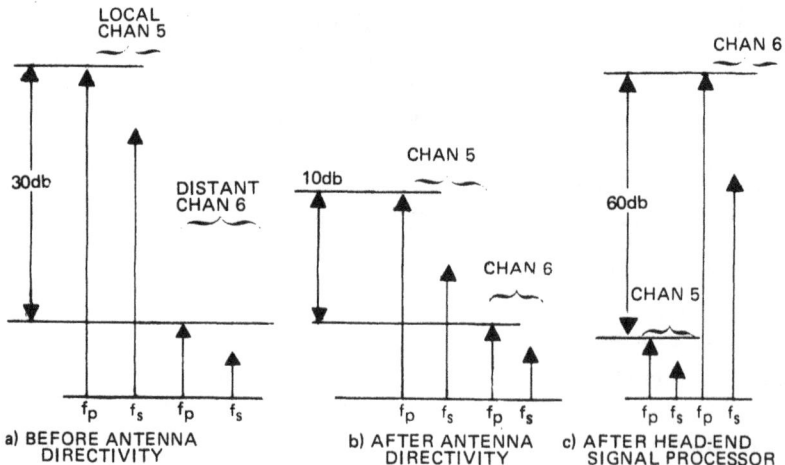

Figure 16.2 Adjacent Channel Filtering from Antenna to Cable Systems
Source: Scientific Atlanta

Figure 16.2 illustrates an example of adjacent channel filtering. The local channel 5 signal denoted by f_p and f_s for picture and sound carrier is shown 30 dB higher than the distant channel 6 before the antenna directivity reduces this ratio to 10 dB at the receiver input. The receiver than provides enough filtering such that the channel 6 output applied to the cable system will be 60 dB greater than the channel 5 output.

16.9 Amplification of the Desired Signal

The head-end signal processor must amplify the signal received from the antenna to a level which will properly load the cable system. This level must be high enough to offset the losses of the multi-channel combining network or signal mixer.

16.10 Automatic Level Control of Picture and Sound Carrier

The head-end signal processor must maintain picture carrier and sound carrier levels at its output at a constant level, regardless of signal level variation from the antenna. The signal level at the antenna terminals will vary from day to night, day to day, and season to season as a result of atmospheric propagation effects and broadcast station variations. The picture carrier is usually maintained at a constant level by an AGC system in the signal processor. The sound carrier is normally separated from the composite RF television signal, and limited or subjected to AGC action in order to maintain a constant level.

16.11 Desired Level Ratio between Picture and Sound Carrier

The head-end signal processor is designed to allow independent control of picture carrier and sound carrier levels. The ratio between these carriers can be set at the head-end. This ratio is normally 15 to 17 dB. This feature is necessary to allow the subscriber TV sets to operate amid adjacent channel input with minimum interference. Television sets are simply not designed for adjacent channel operation.

16.12 Channel Conversion

The head-end signal processor changes the frequency band of an off-the-air signal to an interference-free frequency band for application to the cable. This is necessary to eliminate multi-path problems caused by direct pick-up at the subscribers set.

16.13 Spurious Signals

The spurious signal specification, when listed, refers to spurious responses in TV channels. This usually means that the local oscillator (LO) of the up-converter is only 30 to 40 dB down relative to the picture carrier. This may be satisfactory for most systems, but multi-channel (i.e., greater than twelve) may suffer.

16.4 Down-Converter (Tuner)

The down-converter, or tuner, in the heterodyne may be identical to the turer in the demodulator. The tuner serves exactly the same purpose for both applications. The tuner provides RF amplification, a local oscillator, and a mixer to translate the incoming signal to the IF band. The frequency response of the tuner should be broad enough to pass the incoming channel frequencies (5.25 MHz band) without distortion.

The single-channel crystal-controlled tuner has gained popularity in recent years. For a multi-channel tuner, automatic frequency control is desirable. (Only the single channel tuner will be discussed in detail in this chapter.)

16.15 RF Amplifier

The RF amplifier provides selectivity to the desired channel; i.e., to amplify only the desired channels. The amplifier must have broad enough frequency response so as not to shrink the incoming signal bandwidth. Also, because selective filtering at RF is difficult, the RF amplifier provides very little adjacent channel filtering and, therefore, it must be linear and not generate cross-modulation distortion between adjacent channels and incoming channel signal. Adjacent channel trap circuits in the IF amplifier filter out the adjacent channel signals.

The selectivity of the RF amplifier usually provides sufficient image rejection, and also keeps the local oscillator signal from appearing on the input terminals and radiating back through the antenna. The RF amplifier is normally part of the AGC loop via a delayed AGC voltage.

16.16 IF and Local Oscillator Frequencies

The IF is fixed for all channels and has been chosen to correspond with the standard television IF. In other words, the picture carrier frequency at IF for all incoming channels is 45.75 MHz, and the sound carrier frequency at IF for all incoming channels is 41.25 MHz. The incoming channel has been inverted by the choice of local oscillator frequency in the down-converter, so that the picture carrier frequency is higher than the sound carrier frequency. The video IF amplifier response is shown in Fig. 16.3.

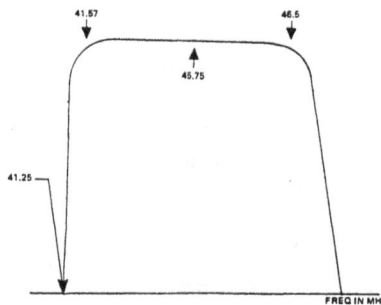

Figure 16.3 Video IF Amplifier Response

Because the sound carrier is separately amplified and limited, the video IF response shows the sound carrier of 41.25 MHz at a very low level (−50 dB desired). To continuously pass the video information, the video IF amplifier response must be constant, or flat, at 4.18 MHz above the carrier (below carrier at IF, or to 41.57 MHz) and 0.75 MHz below the picture carrier (above carrier at IF, or to 46.5 MHz).

The video IF amplifier also provides adjacent channel filtering. The IF response inverts the incoming frequencies, so that the terminology in the figure is correct.

In order to convert all incoming frequencies to IF, the local oscillator frequency must be 45.75 MHz above the incoming picture carrier frequency. This relationship is expressed by the simple equation,

$$F_{IF} = F_{10} - F_{in} = 45.75 \text{ MHz} \tag{16-1}$$

Because the local oscillator is usually a crystal oscillator, the crystals are not normally available above 150 MHz, so the crystal local oscillator frequencies for channels 7 through 13 are therefore doubled. For example, for channel 13, the local oscillator contains a 128.5 MHz crystal. The 128.5 MHz frequency of the oscillator is then multiplied by two, or doubled, to generate the required 257 MHz local oscillator frequency.

If the output channel is to be at the same frequency as the input channel, then the local oscillator in the tuner provides the local oscillator signal for the up-converter. This same result could also be obtained by using the local oscillator signal in the up-converter to drive the mixer in the tuner.

Since the IF frequency is known (45.75 MHz), as is the input or output carrier frequency, the LO frequencies can be calculated directly. For on-channel conversion, local oscillator frequency number one (LO_1) is equal to local oscillator frequency number two (LO_2). This can be proved by considering Eqs. (16-2) and (16-3).

$$F_{IF} = F_{LO_1} - F_{IN} \tag{16-2}$$

$$F_{IF} = F_{LO_2} - F_{OUT} \tag{16-3}$$

$$F_{LO_1} - F_{IN} \text{ and } F_{LO_2} - F_{OUT}$$

then

$$F_{OUT} - F_{IN} = F_{LO_2} - F_{LO_1}$$

If

$$F_{OUT} = F_{IN}$$

then

$$F_{LO_2} = F_{LO_1}$$

16.17 Automatic Gain Control (AGC)

The IF amplifier gain remains constant until the input signal level reaches −14 dBmV. This level is known as the *AGC threshold level* because, for input signals higher than this value, the IF amplifier gain is controlled and its output is held constant. Because of delayed AGC action, the tuner remains at 20 dB until the input signal increases above 10 dBmV. Thus, the delayed AGC threshold is shown as 10 dBmV. Above this input signal level, the tuner gain decreases and the input to the IF amplifier remains constant at 30 dBmV.

The IF amplifier gain is reduced by AGC action over the input signal level range from −14 dBmV to +10 dBmV in order to maintain a constant output level. Over this input level range, the tuner operates at maximum gain and minimum noise figure. When the input level reaches +10 dBmV, the delayed AGC circuit reduces the gain of the tuner, whose output now remains constant at +3 dBmV. At input levels above +10 dBmV, the tuner noise figure does not have a significant effect on the signal-to-noise ratio. If the gain of the tuner were not reduced, then the IF amplifier might be overdriven.

16.18 Sound IF Amplifier-Limiter

In the television converter with IF amplifier, the sound carrier at IF is distinguished from the composite signal and amplified separately. The sound carrier is also amplitude limited, so that a constant sound carrier level may be applied to the cable system. A level control is required to adjust the ratio of video IF carrier to sound IF carrier.

Some general design considerations for the sound IF amplifier-limiter include: minimum gain for a specified amount of limiting at normal operating levels; amplifier bandwidth necessary to keep from adding color sub-carrier sidebands back into the signal path of the combining network; and output signal level required for the correct video carrier to sound carrier ratio.

16.19 FM Signal Characteristics

The sound information of a television signal is transmitted on a frequency modulated (FM) carrier. Minimum amplifier bandwidth is in part determined by the information bandwidth of the FM sound signal. A general expression for an FM signal is

$$e = E_m \sin \left(2 \, F_o t + \frac{Dm}{f_a \, t} \right) \qquad (16\text{-}4)$$

where

D_m = maximum peak deviation

f_a = audio modulating frequency

$m_f = \dfrac{D_m}{f_a} = 5$

because

$$D_m = 75 \text{ kHz and } f_a = 15 \text{ kHz} \tag{16-5}$$

Given that 100% modulation = 5, therefore, bandwidth = 3.2 D_m = 240 kHz; for television sound systems, 100% modulation = 1.667, therefore, $m_f = 25/15$ = 1.667 and bandwidth = 5 x 25 kHz = 125 kHz.

Because of the low probability that the spectral components of the FM signal above 100 kHz will occur, bandwidth of 200 kHz is sufficient. To allow for transmitter and local oscillator tolerances, 100 kHz is the sound IF bandwidth of choice.

16.20 Maximum Bandwidth Considerations

The sound IF amplifier-limiter must amplify the sound signal while providing enough color information rejection in the video passband so that no interference is produced when the sound IF output is combined with the video IF output.

The sound signal is normally operated 15 dB (±2 dB) below the video carrier level.

Figure 16.4 shows the frequency spectrum of the combining network output. The 42.17 MHz which comes through the sound IF amplifier at an arbitrary phase must have a minimum amplitude so as to have a negligible effect on the color signal. As an example, consider 42.17 (e_2) at 1/10 the amplitude (–20 dB) 42.17 (e_1) signal, as given in Fig. 16.4.

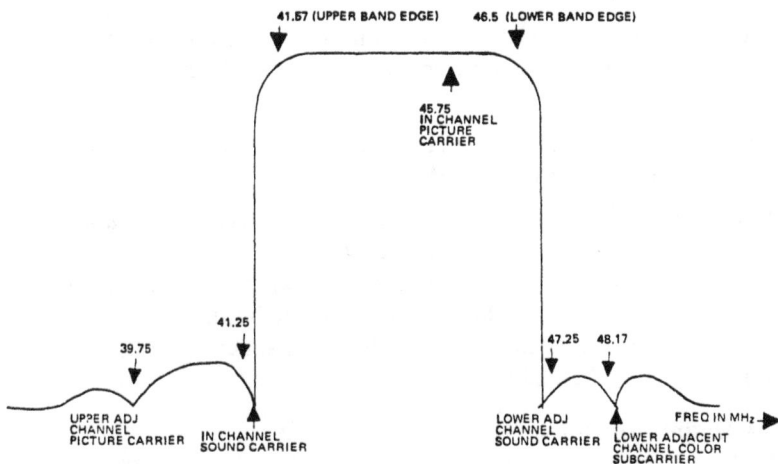

Figure 16.4 Adjacent Channel Filtering
Source: National Cable Television Institute

(ART INSERT FIGURE 16.4) An interfering signal of equal frequency but arbitrary phase and low amplitude would have maximum effect if its phase and amplitude were changing. With minimum rejection of the e_2 signal, there will be little effect on the color picture for two reasons. First, the e_2 signal has minimum rejection, and, second, the phase and amplitude of the interfering signal remains fairly constant.

16.21 Amplifier-Limiter Levels

As an example, consider a design goal for the sound IF amplifier-limiter of 10 dB limiting μV input (–20 dBmV) to the tuner input and a minimum output of 20 dBmV at maximum gain. The output impedance is a nominal 75 ohms at high return loss, providing a good match to the passive combining network.

16.22 Up-Converter and Mixer

The up-converter local oscillator can be identical to the input local oscillator. The input to the output mixer is the constant IF for all input channels and its output is tuned to the required output channel. Be advised that the mixer in the tuner is just the reverse insofar as its input is tuned to the incoming channel frequencies and its output is the same for all incoming frequencies.

The output mixer design can be optimized because the IF amplifier is maintained at a constant level by the AGC amplifier. If the output level control is placed between the IF amplifier and up-converter, the control level range is limited because possible system signal-to-noise ratio effects if the input level to the up-converter is not held above a minimum value.

16.23 IF Conversion

IF frequency conversion to channel 6 output is difficult because the second harmonic of the intermediate frequencies appear directly in the channel 6 band.

The second harmonic of the sound carrier is 82.5 MHz and second harmonic of the color sub-carrier is 84.34 MHz.

Two methods of avoiding this problem are: (1) use of a *double-balanced mixer*, and (2) *double conversion*. The double-balanced mixer balances the second IF harmonic which would otherwise appear at the mixer output. Double conversion avoids this problem entirely by first mixing the IF to a frequency range not harmonically related to channel 6 and then effecting a second mixing to channel 6.

16.24 Output Amplifier

The output amplifier must provide the power gain to amplify the signal to its 54 to 60 dBmV output level. The loss of signal combining networks is inversely proportional to the number of channels to be combined. Therefore, combining 20 channels requires heterodyne converters operating at higher levels than in a 12-channel system.

Because the output stage is operating at high signal levels, transistors similar to those used in CATV repeater amplifiers are also used here to avoid cross-modulation and intermodulation among the picture, sound, and color sub-carriers. The output RF amplifier must be selective in order to filter spurious outputs. The local oscillator at the output mixer is generally 10 to 15 dB greater in amplitude than the IF signal. Considerable filtering must be provided if the up-converter local oscillator is to be more than 60 dB down.

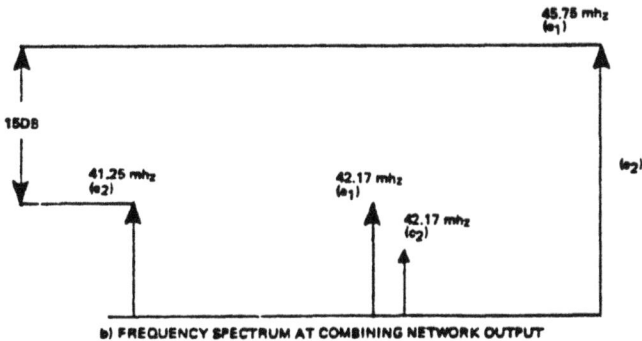

b) FREQUENCY SPECTRUM AT COMBINING NETWORK OUTPUT

Figure 16.5 Sound IF System
Source: National Cable Television Institute

17 Demodutator and Modulator

17.1 Amplitude Modulation Theory: System Considerations

The video signal output of the television camera and the associated synchronizing signals is limited in bandwidth to 4.2 MHz. Because signals in this
frequency range cannot be efficiently transmitted over the airways, the video
signal must be translated to a more efficacious frequency range. One way of
translating the video signal to an efficient transmission frequency range is to
amplitude modulate an RF carrier. Simple double-sideband modulation
results in an RF bandwidth twice that of the video bandwidth. The audio
signal is displaced by 4.5 MHz from the RF carrier and, therefore, requires
more bandwidth, as shown in Fig. 17.1

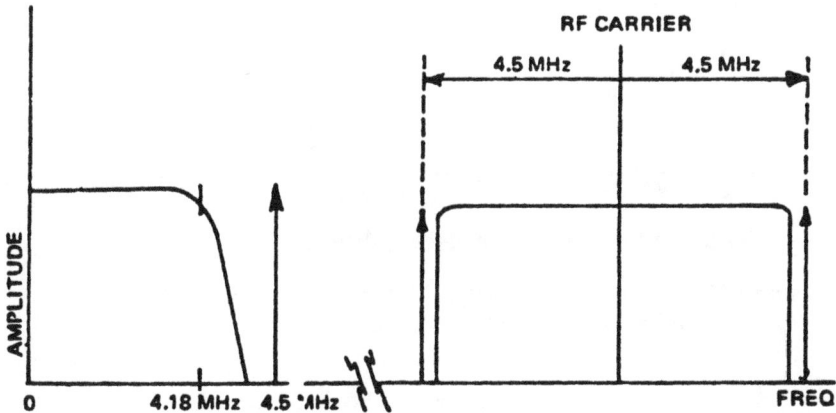

Figure 17.1 Double Sideband Frequency Spectrum

Two of the most important parts of a modulator are the amplitude modulator and the vestigial sideband filter.

The television system signal is restricted to vestigial sideband transmission, not double-sideband, for bandwidth conversion. All of the information necessary to reproduce an excellent television picture and its audio signal is contained in the vestigial signal.

The television RF carrier of a channel is first amplitude modulated and appears as a double-sideband signal. This double-sideband signal is then passed through a filter which produces its vestigial sideband characteristic.

Amplitude modulation often implies a modulating signal varying slowly — for example, the video signal — compared to the modulated signal, the carrier. The amplitude of the carrier signal is varied in direct proportion to the amplitude of the modulating signal. Amplitude modulation results in sidebands spaced above and below the carrier frequency by an amount equal in frequency to the modulating frequency. For example, if a 4.5 MHz sine wave amplitude modulates at 55.25 MHz, the frequency spectrum shown in Fig. 17.2 results.

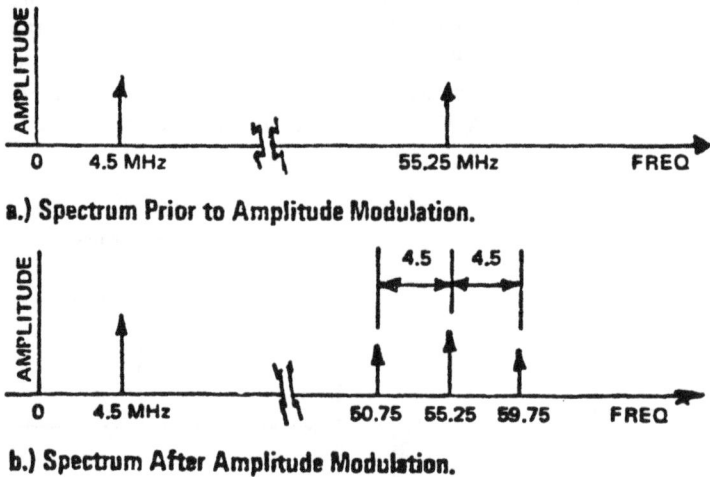

a.) Spectrum Prior to Amplitude Modulation.

b.) Spectrum After Amplitude Modulation.

Figure 17.2 (a) Spectrum Prior to Amplitude Modulation. (b) Spectrum After Amplitude Modulation.

17.2 Mathematical Expressions for Amplitude Modulation

The carrier of a television signal can be considered a sine wave and can be expressed:

$$Vc = A_c \cos W_c t \qquad (17\text{-}1)$$

A_c = Peak voltage amplitude of carrier

$W_c = 2 f_c$ where f_c is the carrier frequency

The example of a single wave modulating signal is often shown in text books, because an illustration of a single sine wave is straightforward and understandable. Most functions can be considered the sum of any sine wave signals varying from a few cycles up to approximately 4.5 MHz.

$$V = A_c (1 + m \cos W_m t) \cos W_c t \qquad (17\text{-}2)$$

$W_m = 2 f_m$ where f_m is the modulating frequency

this equation out gives:

$$V = A_c \cos W_c t + A_c m (\cos W_m t)(\cos W_c t) \qquad (17\text{-}3)$$

     First   Second

The first term of Eq. (17-3) is the carrier frequency, and the second term turns out to be the two sidebands resulting from the amplitude modulation. The second term may be expanded using the trigonometric identity of $\cos A \cos B$ to:

$$\frac{mA_c \cos (W_c - W_m)t}{2} + \frac{mA_c \cos (W_c - W_m)t}{2}$$

Lower Sideband   Upper Sideband

Note that there are now two frequency components, one is f_m higher than the carrier frequency, and the other is f_m lower than the carrier. (See Fig. 17.3.) Note also that the sidebands have an amplitude equal to m times A_c, the carrier amplitude. For 100% modulation, $m = 1.0$, and the sideband has its maximum amplitude equal to ½ of the carrier amplitude. The single sine wave modulated carrier is shown as a function of time in Fig. 17.4.

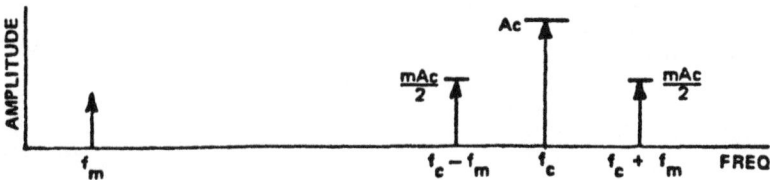

Figure 17.3 Amplitude Modulation Spectrum

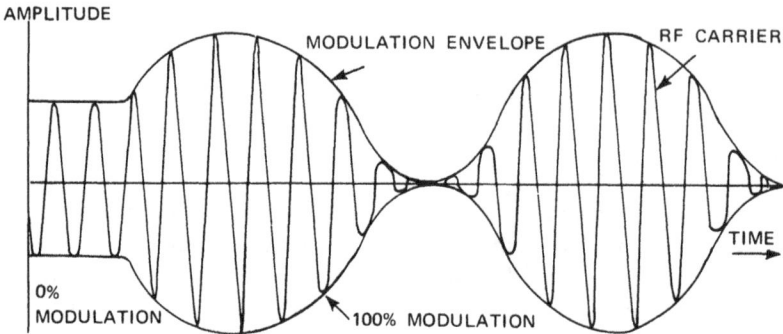

Figure 17.4 Single Sine Wave Modulation
Source: National Cable Television Institute

One difference between the previous example of a single sine wave modulated carrier and the video modulated carrier is that the video modulated carrier is down-modulated. The peak RF carrier level occurs at the horizontal synchronized tip, and the amplitude of the RF carrier decreases as the video signal increases. This type of modulation has the advantage that the peak power requirements of the transmitter are constant, being directly related to the fixed amplitude synchronized signal.

17.3 Amplitude Modulators

The amplitude modulation circuit is probably the most critically important circuit in the modulator. This circuit is where the RF carrier is modulated by the incoming video signal. The output of the amplitude modulation circuit is a double-sideband (i.e., bandwidth equal to twice the video bandwidth) signal centered at the carrier frequency.

An amplitude modulator circuit may be realized by applying an RF carrier to a load resistor through an impedance R_{am} which varies as a function of the video signal amplitude. The amplitude of the output signal e_o will depend upon the magnitude of the input RF carrier and the magnitude of R_{am} which varies directly as the video signal.

$$\frac{E_{out}}{E_{in}} = \frac{R_L}{R_{am} + R_L} \qquad (17\text{-}4)$$

if $\quad R_{am} = R_L$

then $\quad \dfrac{E_{out}}{E_{in}} = R_L / R_{AM}$

17.4 Amplitude Modulation Circuit Distortions

An imperfect clamp circuit is likely to cause synchronized compression (i.e., squashing of the synchronized pulse) and compression of the blacks (low percentage modulation) resulting in differential gain and differential phase. If the frequency response of the video amplifier is not low enough, tilt will occur and excessive tilt can cause loss of vertical synchronization and shading of large-area pictures.

White compression occurs at high modulation percentages. The output does not decrease linearly with respect to the input video and tends to be squashed again. Thusly, white compression occurs. This condition also produces differential gain and differential phase. Let us define these terms:

Differential Gain. The amount of change in the level of the 3.58 MHz color sub-carrier as a function of modulation percentage.

Differential Phase. The amount of change in phase shift of the 3.58 MHz color sub-carrier as a function of modulation percentage.

It logically follows that the most critical conditions for the modulator with respect to these two characteristics are at low and high modulation percentages.

17.5 The Modulator

The modulator provides a means of converting the primary information signals (video or audio) from their original frequencies to the frequency range required for transmission over the airways. The modulator is basically very simple, containing a video amplifier, an amplitude modulation circuit for the video signal, either a 4.5 MHz FM circuit or a 4.5 MHz limiter (for signals fed from a demodulator sub-carrier output), and a linear RF power amplifier. A thorough examination of amplitude modulation theory follows.

17.6 Modulator Theory

A modulator block diagram is shown in Fig. 17.5. Let us visualize segmenting the modulator into video modulation and audio modulation sections. The modulator has two inputs, for video and audio (or a 4.5 MHz FM modulated sub-carrier in lieu of the audio signal). The modulator output is the combination of the video and audio outputs.

17.7 Video Modulation Section

First, let us consider the important specifications of the video modulation section which should appear on the manufacturer's data sheet.

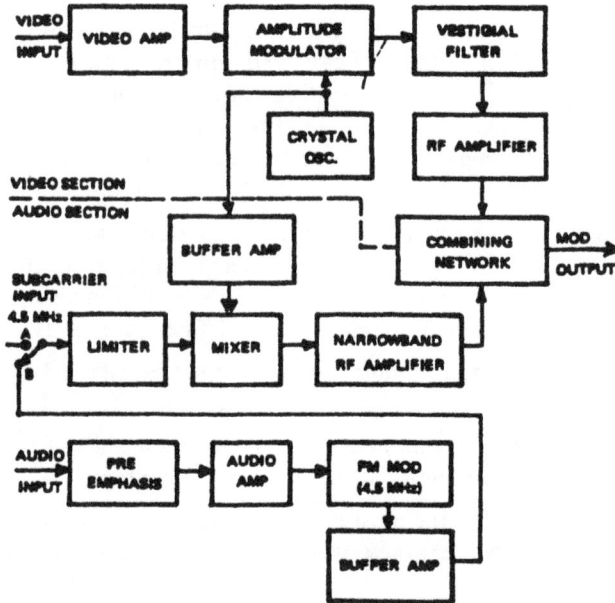

Figure 17.5 Modulator Block Diagram

17.8 Video Input

The input impedance of the video amplifier is normally determined by a 1% resistor shunted across the high input impedance of the amplifier in order to force the match to 75 ohms (variation 2 ohms or less). The input level range is defined by the gain of the video amplifier. The level range specified assures full 87.5% modulation over the video amplifier gain range. The modulation percentage control on the front panel can vary the video amplifier gain.

17.9 Peak White Clip

A modulator equipped with a *peak white clip* circuit insures that the modulation percentage will not exceed an adjustable maximum modulation percentage. Down-modulation is utilized and, therefore, video noise peaks could over-modulate the carrier in the white amplitude area.

17.10 Tilt

The tilt specification may also be referred to as dc restoration, specified over the full modulation range or the white window test. It is a measure of the low frequency response of the video amplifier and the dc restoration circuit. Keep in mind that the RF carrier is down modulated so that the synchronized tips (peak video signal) must be referenced to a solid dc level in the amplitude modulation circuit. This dc reference level must be maintained from maximum to minimum video level.

Modulator Specifications — Video Section

Video Input Impedance	75±2 ohms
Level (for 87.5% modulation)	0.3 to 2.0 volts peak to peak
Peak White Clip (adjustable)	75 to 95%
Tilt (60 cps square wave)	≦1%
Differential Gain (87.5% modulation)	±1.0 dB
Differential Phase (87.5% modulation)	±1.0°
RF Carrier	
Frequency Tolerance	±.01%
Frequency Stability (0°C to +70°C)	±.005%
RF Output	
Level (adjustable)	+45 dBmV to +60 dBmV
Impedance (RL referred to 75 ohms)	18 dB
Spurious Outputs	≧ −60 dB
Frequency Response (Passband)	±0.25 dB
AM Modulation on RF Sound Carrier	0.1% maximum

17.11 Differential Gain and Differential Phase

Differential gain and differential phase can occur in the video amplifier, but proper design can virtually eliminate both in the video modulation circuit. The amplitude modulation circuit is normally the only circuit of the modulator where differential gain and differential phase could occur. These two types of distortion are sometimes specified at 10% and 90% *average picture level*

(APL). A measurement of the differential gain and differential phase at 10% APL and 90% APL ensures that the specification is met at low and high percentage modulation, respectively.

17.12 RF Carrier Frequency

The RF carrier frequency is normally determined by a crystal controlled oscillator. Two tolerances may then be specified: the absolute accuracy of the crystal frequency and the temperature stability of the crystal frequency.

17.13 RF Output

Several output specifications are necessary to describe the quality of the output signal. Not all of the specifications discussed may appear on manufacturers' data sheets. First, the output level of the video carrier should have an adjustment range. (After the picture carrier to audio carrier level ratio is set, the video carrier should be adjusted to the combined signal levels.) The output return loss usually applies to the passband only, but the frequency response should apply from the video input to the RF output. The frequency band should be specified, such as from 0.75 MHz below to 4.5 MHz above carrier. The last item mentioned is amplitude modulation on the RF sound carrier. This specification is a measure of the amplitude linearity of the RF amplifier and as such is really a form of cross-modulation.

17.14 Demodulator Theory of Operation

The discussion of the demodulator is presented here in the context of demodulator theory of operation, an approach similar to that taken for modulator theory above.

17.15 Input Level

The input level range is limited at the low end by the maximum gain and noise figure of the RF amplifier and the gain of the IF and AGC amplifiers. The minimum useable signal should be within the AGC range of the demodulator so that the video output will be maintained at a constant level. The signal-to-noise ratio can be calculated when referred to 75 ohms, and the signal will be quite noisy below about –6 dBmV.

The maximum signal is limited to a level which will not overdrive the RF amplifier and the mixer. If these circuits are overdriven, various spurious responses will be produced. The maximum input level may be limited further by the presence of a strong adjacent channel carrier, causing cross-modulation. A high Q trap may be required for some applications to reduce the level of the adjacent carrier signal.

For most applications, a good antenna location should limit input signal level variations to $<$ 10 dB about the nominal value of 0.1.

17.16 Noise Figure

Demodulator noise figure should be as low as possible. If the noise figure is >
6 dB, an RF amplifier should be used for input signal levels below about –6
dBmV. Of course, the RF amplifier noise figure should be on the order of 4
dB.

17.17 Selectivity

This specification refers to adjacent channel traps located in the IF amplifier.
These traps reject the adjacent channel carriers and are especially necessary
for application where the desired channel is not as strong as the adjacent
channel. One characteristic to beware is the temperature stability of these
traps. The trap frequencies may drift with time and temperature because of
their sharpness.

17.18 Input Impedance

It is imperative that the input impedance be well matched to 75 ohms. If the
demodulator input is connected from the antenna to the input via a low loss
cable (this is the usual case), the probability of reflections and ghosts is high
unless the input match is good. The match should remain constant over the
AGC range as well. This feature is sometimes not available, and, if there is
enough signal available, a match may be forced by connecting a 6 dB pad
between the demodulator input and the antenna down-lead.

17.19 AGC Range

A demodulator with a 60 dB AGC range should contain delayed AGC.
Delayed AGC allows the RF amplifier to operate at maximum gain until the
input signal reaches a level where the signal-to-noise ratio will not suffer by a
reduction of gain in the RF amplifier. The RF amplifier may have an AGC
range of 20 to 30 dB and the IF amplifier has an AGC range of 30 to 40 dB.

17.20 Differential Gain and Phase

Differential gain and differential phase are difficult to minimize in the *video
detector* following the IF amplifier. Indeed, this is where these two characteristics are most difficult of all. The video detector is the only circuit area in a
properly designed demodulator where these distortions should occur.

17.21 Video

The video output level is normally made adjustable over a convenient range
which applies only when the system is in the AGC mode of operation. The
video output match is normally forced by a 73 ohm *precision resistor* placed
in series to the output from an *emitter follower*, or by the use of 75 ohm
collector (or plate) *load.*

17.22 Audio Level

The 4.5 MHz sub-carrier is normally circumvented at the video detector. It is then routed to a discriminator which results in an audio signal, which is then amplified to the desired level.

We have discussed the demodulator and modulator specifications in some detail. This chapter concludes with a closer look at a few items.

17.23 Tuner

The tuner consists of an RF amplifier, local oscillator, and mixer. This tuner may be either crystal controlled or an all-bands LO tuner. The all-band tuner is prone to drift, but provides the flexibility of all-band tuning. It is possible to incorporate an automatic frequency control (AFC) circuit in the tuner, similar to that used in the *channel commander* heterodyne unit. The mixer output sets the video carrier at 45.75 MHz, the audio carrier at 41.25 MHz, and the color sub-carrier at 42.17 MHz.

It is important to note the video carrier is above the audio carrier at the mixer output, which is inserted from the RF input. The reason this happens is that the local oscillator is operated above the incoming RF carrier and the mixer output is the difference frequency.

As an example, we take the channel 5 input, where

Local Oscillator = 123.00 MHz
RF Video Carrier = –77.25 MHz
IF Video Carrier, or Difference = 45.75 MHz

Local Oscillator = 123.00 MHz
RF Audio Carrier = –81.75 MHz
IF Audio Carrier, or Difference = 41.25 MHz

17.24 IF Amplifier Response

The IF amplifier response is worth reviewing here. Because the television signal is transmitted vestigial sideband (the low frequencies are double-sideband), more energy is present at those frequencies which are transmitted double-sideband. A flat IF amplifier response across the entire 5 MHz channel would result in the response shown in Fig. 17.6 and the IF responses shown in Fig. 17.7.

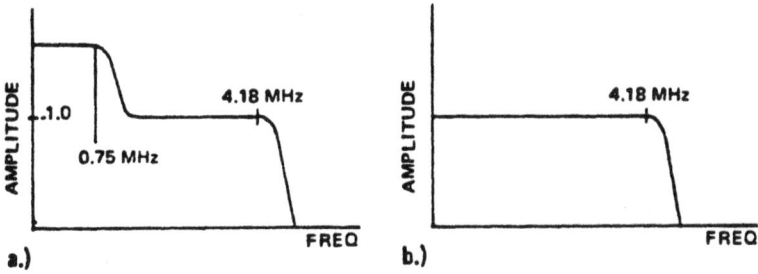

Figure 17.6 Video Response. (a) Video Response, 5MHz IF Bandwidth; (b) Video Response, Ideal IF Response

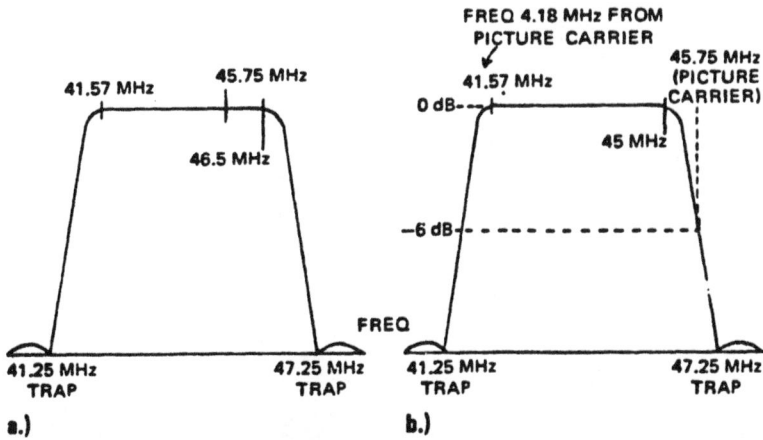

Figure 17.7 IF Amplifier Response. (a) Broad Response, Heterodyne IF (b) Ideal IF Response

18 Preamplifiers and Filters

Filter theory is basic to all communications systems. Information, TV signal, voice signal, data, *et cetera*, are transmitted in a given frequency bandwidth over or through some medium to a receiving terminal. The receiving terminal must process the transmitted signal and present it in such a way that none of the transmitted information is lost. The receiving terminal must provide frequency selectivity, or some other type of filtering, so that unwanted signals do not interfere with the desired signal. Unwanted signals may be random noise, man-made noise, or signals transmitted from another source.

As an example of a filter application: channel 6 is a local channel easily received at the CATV head-end site. Channel 5 is remotely located, but is also available to the system. The channel 5 antenna is made as directive as possible, but some filtering is necessary to prevent overloading the channel 5 receiver. This filtering may take the form of a bandpass filter and a high Q trap with frequency response.

18.1 Properties of Filters

A *filter* is a circuit consisting of impedances grouped together in such a way as to have a definite frequency characteristic.

The amplitude response is unity in the passband, there is no signal attenuation, the phase response is linear, and the time delay response is constant. A linear phase response results in a constant time delay. Specifying the amplitude response and either the phase response or time delay response defines the ideal filter. A signal applied to the filter input will appear at the output, unchanged in amplitude and delayed in time by an amount t_d. The dashed lines in Fig. 18.1 correspond to a practical filter response.

18.2 Filter Response Terminology

An infinite number of elements, inductors, and capacitors is required to realize the ideal filter amplitude and phase response. Obviously, therefore, the filter designer must use some approximates in designing a filter. More elements and greater circuit complexity are needed as the approximation of the ideal filter response is approached. Two approximation techniques are the maximally flat amplitude, or *Butterworth response*, and *Chebyschev response*. Each of these responses is an attempt to achieve the best possible response given the smallest number of elements. First, let us briefly explain the maximally flat Butterworth response.

The Butterworth approximation applies to bandpass, high-pass, and low-pass filters. The example of Fig. 18.2 considers low-pass filters.

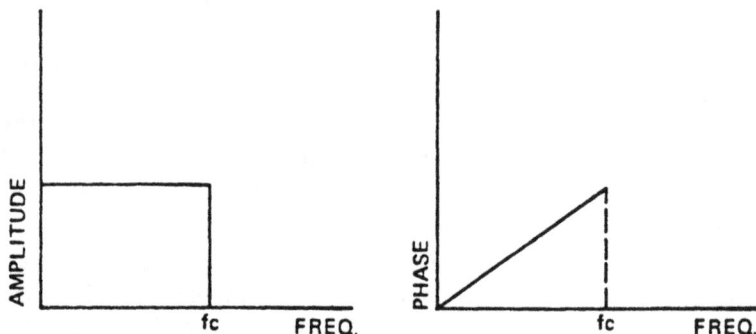

Figure 18.1 Ideal Filter Responses

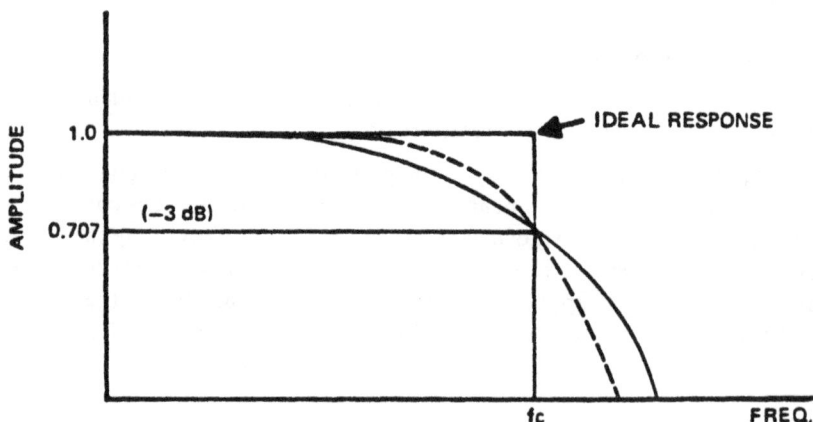

Figure 18.2 Maximally Flat Characteristic

The ideal response approximation is good at low frequencies (for a bandpass filter it would be good near the center frequency; a high-pass filter at higher frequencies), but is poor at low frequencies. The dashed line indicates a filter with more elements, which renders a better approximation than one with fewer (the solid curve). The pass-band response is smooth and has no ripples.

Chebyschev response is explained by comparison with the Butterworth. The maximally flat approximation lumped all of its errors with respect to the ideal response at high frequencies or near the cut-off frequency. The Chebyschev, or equal-ripple, approximation distributes the error throughout the pass-band. The Chebyschev response filter provides a faster rate of cut-off (steeper skirt) compared to the maximally flat response filter.

18.3 Preamplifier

Preamplifiers are normally used in circumstances where the received signal level is too low to provide a quality picture. Most preamplifiers have input and output impedance of 75 ohms, so that they can be used directly with regular coaxial cables without impedance matching devices. The preamplifier is normally placed very close to the antenna, because the signal-to-noise ratio of any system is usually controlled by the first amplifier in the system. Preamplifiers usually have a noise figure better than 4 dB. It should be noted here that preamplifiers will only operate effectively in circumstances where the signal level is only a few dB too low. In more extreme cases it may be necessary to go to a higher gain antenna and preamplifier.

18.4 Typical Filter Specification

A typical filter data sheet might read as follows:

passband	76 to 82 MHz
response flatness	1 dB peak-to-valley
3 dB bandwidth	6.5 MHz
insertion loss	1.5 dB
impedance	75 ohms ($VSWR$ 1.3 at center frequency)
selectivity	attenuation 40 dB at frequencies 6 Mhz beyond passband

Similarities between low-pass, high-pass, and bandpass filters should be noted here. The bandpass filter has been discussed in detail, and the description of any of the three types of filters can be made using precisely the same terms. Each type of filter has cut-off frequencies, skirt steepness (selectivity), insertion loss, response flatness, bandwidth (bandpass), impedance, *et cetera*.

18.5 Insertion Loss

The signal generator is a heterodyne unit required to drive a CATV system cable. The cable impedance is R_L and the heterodyne output impedance is R_g. Delivery of maximum power to R_L (the cable) by generator E_g requires that $R_g = R_L$.

$$I = \frac{E_g}{R_g + R_L} \text{ and } E_o = IR_L \qquad\qquad (18\text{-}1)$$

so that

$$E_o = E_g \frac{R_L}{(R_L + R_g)}$$

$$E_o = E_g \frac{1}{2}$$

(since $R_L = R_g$)

For maximum power transfer, the output voltage E_o is one-half of the generator voltage E_g. This apparent loss of one-half of the generator voltage is not insertion loss, but occurs when the maximum power is transferred from the heterodyne unit to the cable. A filter has been inserted between the generator and load. This is usually necessary to ensure that only the desired frequencies from the heterodyne unit are applied to the cable. If the filter has 6 dB insertion loss, the voltage across the load R_L, in terms of the generator voltage E_g is

$$E_o = \frac{E_g}{2} \text{ (No Filter)}$$

$$E_o = \left(\frac{E_g}{2}\right)\left(\frac{1}{2}\right)$$

18.6 High Q Trap or Brand Elimination

The high Q trap type of filter may be considered the inverse of a bandpass filter. Whereas a bandpass filter allows a band of frequencies to pass through to the load, a band elimination filter allows all frequencies to pass except for those in the stop-band. A high Q trap is a special case of the band elimination filter, which has a very high insertion loss over a very narrow band of frequencies.

High Q traps are often used in CATV head-ends to filter adjacent channel picture or sound carriers. High Q traps usually have only a few elements and are much less complicated than other band elimination filters.

19 Fundamentals of Transmission Lines

A transmission line is a combination of electrical conductors used to transport some form of electrical energy from the generating source to a load. Any pair of wires can transport electrical energy, but special electromagnetic requirements must be satisfied to do the job efficiently. For CATV purposes, the antenna system is usually the generating source and television receivers in subscribers' homes comprise the load. Some type of transmission line must be used to efficiently carry the RF energy of television signals between these two points.

19.1 Balanced Lines

A *balanced transmission line* consists of a symmetrical arrangement of two identical conductors (wire) separated at some constant specific distance by an insulating and spacing material (usually polyethylene). The twin lead used by conventional television sets between the receiver and its antenna is such a balanced transmission line.

When a transmission line is said to be balanced, it means that equal currents are flowing through the two conductors in opposite directions. Thus, the electric and magnetic fields generated by the current in the conductors are equal in magnitude and opposite in direction, and therefore cancel. This cancellation prevents radiation losses from the transmission line and, consequently, prevents the transmission line from causing interference in nearby receivers and other electronic equipment.

The major disadvantage of a balanced transmission line lies in the fact that it can absorb electromagnetic radiation from external generating sources such as power lines, vehicle ignition systems, and radio transmitters. In a CATV system this represents additional noise and spurious signals in the system which would degrade system performance markedly.

19.2 Coaxial Lines

A *coaxial line* consists of two, non-identical, concentric conductors separated by an insulating material. The inner conductor is a solid or stranded wire. The outer conductor, which completely surrounds the inner conductor, is usually a woven braid. If the center conductor is connected to ground, then the outer conductor acts as a shield for the center conductor, thus eliminating any external noise and spurious signals which would degrade performance.

This coaxial arrangement of conductors results in an unbalanced transmission line. Electric and magnetic fields created by current flowing in the outer conductor cannot be cancelled by electric and magnetic fields created by current in the center conductor. (The outer conductor acts as a shield preventing the escape of fields from the center conductor.) This results in radiation from any currents flowing in the outer conductor (shield). If radiation loss is a serious problem, then some external method (many are available) must be used to create an effective balance in the transmission line. (This transmission line balance is, in effect, cancellation of radiating electric and magnetic fields.)

In CATV systems, coaxial cable of various types is used as the transmission line from the antenna and head-end location to a subscriber's home. Because only very low levels of signal current are being carried by the transmission line, radiation loss from the line is not a problem and the unbalanced condition which exists need not be corrected. Coaxial cable is used instead of twin lead (balanced line) because coaxial cable will greatly reduce radiation pickup from any internal radiation sources.

19.3 Characteristic Impedance

All transmission lines are alike in that there are two conductors in the line which run parallel to each other and which are always separated at some fixed distance by an insulating material. A capacitor is nothing more than two parallel conductors separated by an insulating material. Therefore, there is a certain amount of capacitance between the two conductors in a transmission line. Also, because current flowing through a conductor generates a magnetic field around the conductor, which is exactly what occurs in an inductor, in which case each of the conductors in a transmission line appear to have some inductance.

Unlike a conventional circuit where capacitance and inductance are "lumped," capacitance and inductance in a transmission line are "distributed" along the line. This distribution is usually broken down into 1′ sections for ease of analysis.

Any transmission line can be represented by an amount of series inductance per foot between the conductors. This combination of series inductance and shunt capacitance in each section of transmission line forms what is known as the *characteristic impedance*, Z_o, of the line. Z_o is purely resistive and, therefore, consistent for any given transmission line. Impedance is normally a function of frequency; this is not so with the characteristic impedance of an RF transmission line. The characteristic impedance of a transmission line is a fixed quantity of ohms, determined strictly by the physical dimensions of the line. A signal voltage applied to an actual circuit, comprised of actual capacitors and inductors, "sees" the total impedance of the whole circuit (i.e., the impedance varies with signal frequency). A signal voltage applied to a long transmission line "sees" only the characteristic impedance, Z_o, at the input to the line. The characteristic impedance, Z_o, is simply the square root of the ratio of inductance (in henries per foot) to capacitance (in farads per foot) and is, therefore, purely resistive and a constant (i.e., not frequency dependent).

For CATV purposes, the characteristic impedance of a transmission line must, obviously, be 75 ohms. The characteristic impedance of a transmission line has nothing to do with signal attenuation in the transmission line. Characteristic impedance only yields load impedance or input impedance information, so that efficient signal power transfer can be obtained.

Attenuation in a transmission line, such as in the coaxial cable between CATV amplifiers, is caused by the distributed inductance and capacitance in the transmission line. Attenuation in coaxial cable is frequency dependent. As frequency increases, the inductive reactance of the series inductance increases (i.e., tends to block signal), and the capacitive reactance of the shunt capacitance decreases (i.e., tends to short out the signal). This explains why attenuation in a CATV coaxial cable is stated in dB per foot, and why attenuation per foot increases with frequency (the course of "tilt" in a system).

19.4 Velocity of Propagation

The electrical wavelength of any sinusoidal wave is the physical distance traveled by an increment of electromagnetic energy during the time taken to complete one sinusoidal cycle. As frequency (cycles per second) increases, the time taken to complete one cycle decreases and, therefore, the electrical wavelength decreases. Thus, electrical wavelength (measured in feet) of a signal equals the velocity of the energy (feet per second) divided by the signal frequency (cycles per second). The velocity at which the energy travels is known as the *velocity of propagation*.

In free space, the velocity of propagation for electromagnetic energy is the speed of light (usually given as *c*). In all other transmitting media, the velocity of propagation is less than the speed of light. Clearly, as the velocity of propagation decreases, the electrical wavelength decreases.

The velocity of propagation (*VP*) varies for different types of transmission lines. *VP* is expressed as a percentage of the speed of light. Thus, for free space, *VP* is 100%, *VP* for twin lead is considerably less, and *VP* for coaxial cable is smaller still. When constructing an actual transmission line, it is often necessary to know the electrical wavelength in the type of transmission line used, so that the transmission line can be tailored to fulfill requirements and prevent certain problems such as *periodicity*.

19.5 Periodicity

If, in any CATV transmission line, some type of discontinuity appears at regular intervals, a condition known as periodicity arises. A discontinuity can arise in a cable if the cable machinery causes a slight compression in the cable during manufacture. Because the dimensions of the cable have been changed at that point, there is a corresponding change of characteristic impedance. A discontinuity can also occur whenever an amplifier is inserted into the transmission line or whenever a tap is inserted. Thus, any time there is an interruption of any sort in the transmission line, a discontinuity, or characteristic impedance change (of some magnitude), occurs at the point of interruption. Because of sudden change in characteristic impedance, some of the signal power in the transmission line is reflected. Reflected power creates *standing waves* along the transmission line.

If the discontinuities occur at regularly spaced intervals, the distance between discontinuities is a half-wavelength for some frequency. Thus, for this frequency and for all integral (1, 2, 3, 4, ...) multiples of this frequency, the standing waves caused by reflected power are in phase, and therefore add. The result is that most of the power at this frequency is reflected and, therefore, any signals at this frequency are greatly attenuated. If, for example, these discontinuities occurred at a distance which was one-half of an electrical wavelength for television channel 6, then channel 6 could "mysteriously" disappear from a CATV system even before reaching the first subscriber location.

Periodicity can be prevented by careful inspection of all cable before installation, and by locating amplifiers and taps at random (rather than regular) distances from each other.

19.6 Frequency Response

Frequency response is a term used to describe how a device operates as the frequency is varied. The device can be a length of cable, a passive splitter, a directional coupler, or an amplifier. Illustrations of frequency response for flat loss, cable, and two amplifiers are shown in Fig. 19.1, 19.2, 19.3, and 19.4, respectively. The input level is at different channel frequencies and the output signal levels are shown in each case. We can then see the loss (or gain) may be different at each of the various frequencies. In other words, the frequency response is the gain or loss plotted for each frequency over some frequency range or bandwidth. A curve may then be drawn through each discrete frequency point which will show the response over the entire frequency band being considered. Instead of using discrete frequencies, measuring the loss at each frequency, and then plotting a curve, we may use a *sweep generator*. The sweep generator output is a constant voltage that changes frequency, or "sweeps", across a given frequency range. By using a detector and an oscilloscope we can measure and display the amount of gain or loss at every frequency generated by the sweep generator. A test set-up to measure frequency response is shown in the diagram below.

19.7 Cable Loss versus Frequency

In this section we discuss how the unity gain concept can be applied to the cable loss, and how the cable loss varies with the frequency of the signal being transmitted, as shown in Fig. 19.5.

Consider that the cable is made up of small pieces, then we can look at one of those pieces and see the effects of loss in the cable as the frequency changes. We see that each length of cable has a series inductance and a short (parallel) capacitance associated with it. Therefore, as the frequency increases, the series inductance increases the series loss, and the unit capacitance increases signal leakage to the coaxial outer conductor.

Therefore, Fig. 19.6 shows that the loss at channel 2 might be given for a typical piece of cable at 10 dB and, therefore the loss for that same piece of cable at channel 13 would be 20 dB. We then see that the loss in the cable varies with frequency as some logarithmic curve. The attenuation varies approximately as the square root of the ratio of the frequencies involved. In the example of channel 2 and channel 13 the loss would be proportionate to

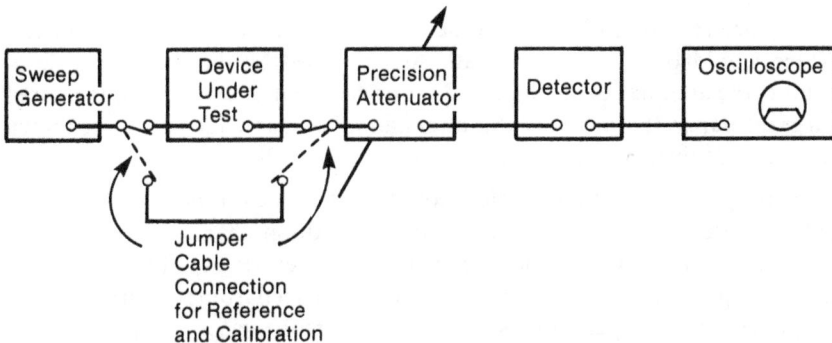

Source: Warner-Amex

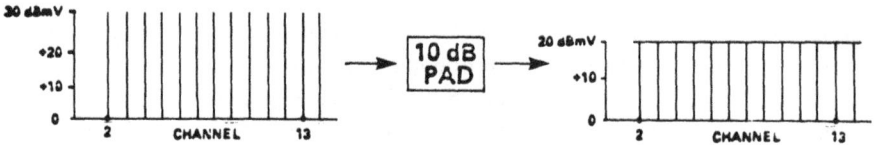

Figure 19.1 Frequency Response/Flat Loss
Source: Texscan Corp.

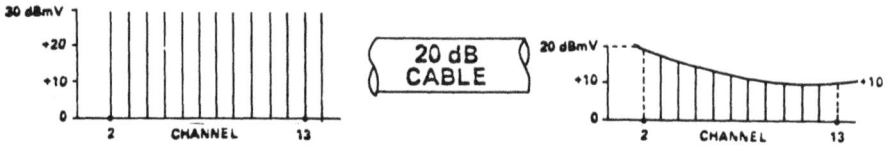

Figure 19.2 Frequency Response/20 dB Cable
Source: Texscan Corp.

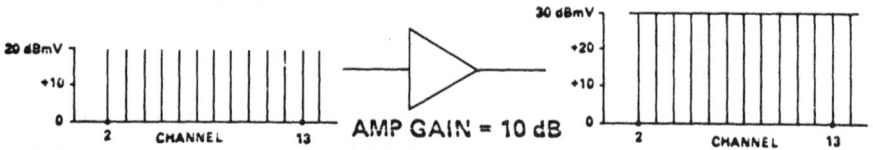

Figure 19.3 Frequency Response/AMP Gain = 10 dB
Source: Texscan Corp.

Figure 19.4 Frequency Response/AMP Gain = 10dB
Source: Texscan Corp.

the square root of 54 MHz (channel 2) divided by 216 MHz (channel 13), which would be equal to one-half. As expressed by the common rule of thumb, the attenuation at channel 2 is one-half that at channel 13. If plotted on graph paper, the curve of cable attenuation *versus* frequency can be shown as a straight line, as in Fig. 19.7.

As noted previously, the amplifier should compensate for the exact loss in the cable. Therefore, we can draw a line from +10 dB at channel 2 to +20 dB at channel 13 and show that the amplifier gain will either exactly compensate, and be opposite to, the loss in the cable. It can be seen from the curve that the amplifier gain is sloped; that is, it varies with frequency.

Should a flat gain amplifier (one having equal gain at all frequencies) be used, it would be necessary to provide equalization to adjust the response of the amplifier to the required slope that is equal and opposite the cable. The gain of a flat amplifier is shown by the dotted lines in Fig. 19.7. In order to compensate this curve to the cable, an equalizer is used to attenuate the low end frequencies and effectively eliminate the shaded area shown in Fig. 19.7. This has the net effect of making the amplifier gain equal and opposite the cable loss.

It also, however, has some disadvantages:

1. In attenuating low frequency signals the carrier-to-noise ratio at these low frequencies is badly degraded. In our example, the signal at channel 2 is reduced by 10 dB. Therefore, the carrier-to-noise ratio at channel 2 would be 10 dB worse than an equivalently sloped gain amplifier.

2. In designing an equalizer circuit, the high end roll-off is rather sharp. When equalizers are cascaded, this roll-off tends to shrink and results in bandwidth compression. Thus, it may be difficult to maintain high end frequency response in a long cascade of amplifiers.

3. Because equalizers consist of discrete components, and all are made essentially the same, they tend to have peaks and valleys at the same frequencies. When cascaded they add at the various frequencies and produce a *signature*, or increased peak to valley deviation, in the response curve of the system.

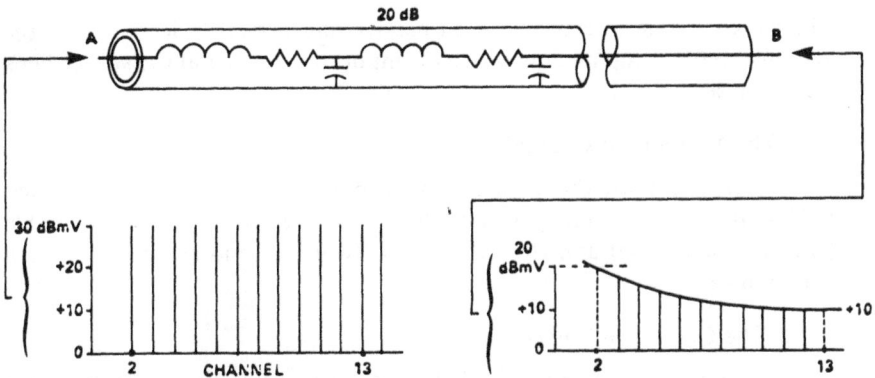

Figure 19.5 Cable Loss & Amplifier Gain vs. Frequency
Source: Texscan Corp.

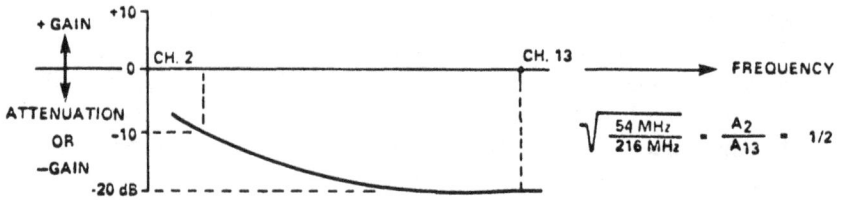

Figure 19.6 Cable Loss & Amplifier Gain vs. Frequency
Source: Texscan Corp.

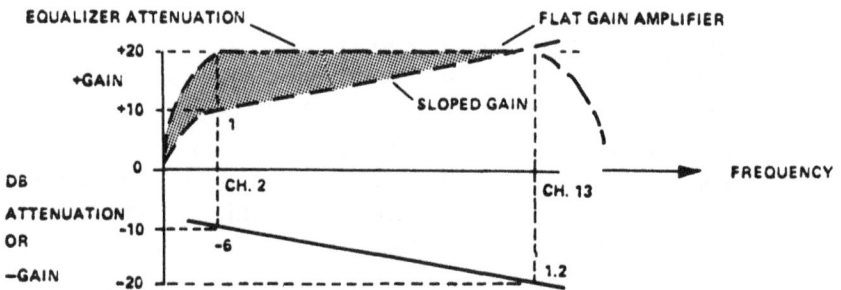

Figure 19.7 Cable Loss & Amplifier Gain vs. Frequency
Source: Texscan Corp.

19.8 Cable Fundamentals

The coaxial cable interconnects all passive and active devices used in the system, from the antenna output to the input transformer at the subscriber's television set.

19.9 The Meaning of Coaxial

Concentric circles are those which have a common axis. Coaxial, when used to describe a cable, simply means that the center conductor of the cable (which is one circle) and the cable shield (which is another circle) have a common axis.

19.10 The Cable as a Power Carrier

A major function of the coaxial cable is to carry power to the active electronic components throughout the system. The efficiency of the cable in carrying power from source to load is determined by the amount of resistance in the cable and the amount of current drawn by the power load.

The comparative level of cable resistance to current flow is a function of the cross-section of the conductors involved. In a coaxial cable, these conductors are the center conductor and the outer shield. The outer shield has a much greater cross-section area than the center conductor. Even though the shield material is aluminum (which is not as good a conductor as the copper used in the center conductor), the size of the shield cross section itself makes its resistance to current flow negligible as compared to the resistance of the center conductor cross section.

To measure the resistance of a piece of cable at dc, interconnect the shield and the center conductor on one end of a 100' section of coaxial cable, then connect an ohm meter between the shield and center conductor at the outer end of the cable section.

The amount of dc resistance measured by the ohm meter is termed the *loop resistance* of 100' of the cable. It is important to remember that the greatest portion of this resistance is supplied by the center conductor. Therefore, the larger this center conductor, the greater the cross section will be and the less the loop resistance or dc loss of the cable will be.

19.11 Loop Resistance in Cable Systems

Cable systems are powered by ac and the nominal frequency is 60 cycles per second. The pictorial representation is a sine wave in which the peaks occur 60 times per second. At this frequency, there is very little difference between the loop resistance in a cable measured at dc and the same resistance measured at 60 cycle ac. Therefore, we can use the loop resistance measured at dc in determining the voltage drop from a power source to any power load because this drop is a function of the current drawn multiplied by the loop resistance.

19.12 The Cable as a Frequency Carrier

As the repetition of the sine wave peaks increases, consider the cable loss is no longer similar to the value at dc. The nature of coaxial cable construction creates characteristics which become controlling factors as frequency increases. The outer shield and the center conductor form capacitance, and both the center conductor and outer conductor are wires which have an element of inductance. Capacitance and inductance are each factors which must be considered as frequency increases.

19.13 The Function of the Outer Shield

In coaxial transmission theory, the outer shield always represents ground to the RF information on the inside of the cable.

19.14 The Cable as a Capacitor

Because the outer shield is at ground potential, the center conductor must be at some potential other than ground. Essentially, the components of a cable are a capacitor with the shield at ground potential, the center conductor at some other potential, and the two separated by a dielectric. See Fig. 19.8 and Eq. (19-1).

When we charge this capacitor from some voltage source, remove the source, and check the voltage after the removal, there will be some voltage remaining, but a lesser quantity than that with which we started. This indicates that there was some leakage between the center conductor and the shield. This leakage is called *dielectric loss*.

19.15 Cable Losses due to Temperature Change

Change in atmospheric temperature is a major cause of signal variation in CATV systems. These temperatures greatly affect losses in the cable. The primary influence of temperature is on the dc resistance of the cable and the dielectric losses. Increases in temperature, like increases in frequency, increase both conductor and dielectric losses in the cable, while decreases in temperature, conversely decrease both of these losses.

$$R_{AC} = Z = R_{DC} + X_L + X_C$$

$$= R_{DC} + \omega L + \frac{1}{\omega C}$$

$$= R_{DC} + 2\pi fL + \frac{1}{2\pi fC} \tag{19-1}$$

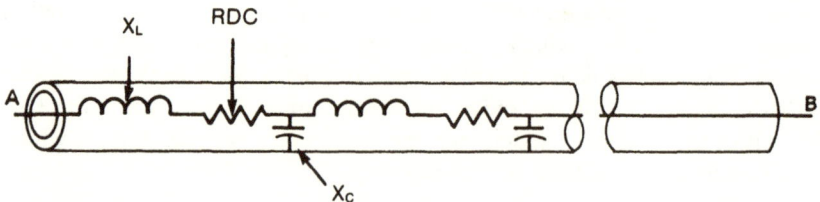

Figure 19.8 The Components of a Cable as a Capacitor
Source: Texscan Corp.

19.16 Basic Rule of Cable

There is a basic rule of cable which states that the cable attenuation caused by frequency variations will follow the square root of the frequency. The relationship between the highest and lowest frequencies carried by the cable is, therefore, constant at any point on the cable.

19.17 Cable as an Inductance

The conventional symbol for an inductance is a coil. Were we to transform that coil into a straight line, it still would have inductance because of the flow of current through it. In the cable, therefore, we have such a straight line in the form of the center conductor. Moving from point A to point B, the center conductor is a *series inductance*, but it is not a *single* series inductance and, therefore, in any transmission line going from point A to B, there will be an infinite number of inductances. Thus, the center conductor and outside shield, with the dielectric material between them, will form a capacitor. This capacitor is a *shunting capacitor* from center conductor to ground. Hence, we can add a series of shunting capacitors to the electrical equivalent of a piece of coaxial cable, so that our representative picture becomes a capacitor at one end of the cable followed by an inductance, another shunting capacitor, another inductance, and so on to the shunting capacitor at the other end of the cable.

19.18 Cable Impedance

The impedance of the cable, as represented by Z, is equal to the square root of the inductance of the cable divided by its capacity. In the CATV industry, this impedance is always 75 ohms. Therefore, the impedance of the cable is purely a function of the center conductor diameter and the spacing from the center conductor to the outside shield, with whatever dielectric in between. The *characteristic impedance* of coaxial cable can be defined in the equation in Fig. 19.9. Because these parameters are entirely dependent upon physical construction, coaxial cable impedance can also be expressed in the equation and diagram in Fig. 19.10. As a result, coaxial cable of 75 ohm impedance may have various center conductor and outer diameters.

19.19 The Importance of Good Cable Installation Practices

It is critically important to exercise the utmost care in the construction of a CATV system. As an example, during the course of construction, if cable is laid along the ground and a moving vehicle runs over the cable, it will flatten the bottom of the cable, whereupon the space from center conductor to outer shield will be changed, consequently changing C without changing L. Therefore, the 75 ohm impedance is changed.

$$Z_o = \sqrt{\frac{L}{C}}$$

Where

L = Inductance in Henrys/unit length

C = Capacitance in Farads/unit length

Figure 19.9 Characteristic Impedance of Coaxial Cable
Source: Texscan Corp.

$$Z_o = \frac{138}{\sqrt{E}} \log_{10} \frac{D}{d}$$

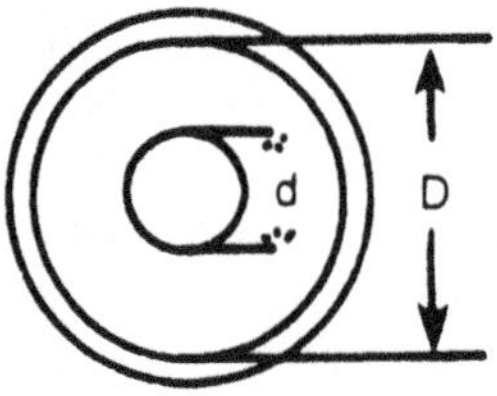

Where

E = Dielectric coefficient

D = Inside diameter of the outer conductor

d = Outside diameter of the inner conductor

Figure 19.10 Impedance Parameters of Coaxial Cable
Source: Texscan Corp.

Alternatively, we may "kink" the cable, in which case the result will be the same because the distance from center conductor to outer shield will be reduced and the capacitance will change without any change in L. Therefore, the impedance is changed.

19.20 The Effect of Moisture in a Cable

The presence of moisture in the cable is a problem that plagues the entire CATV industry. The presence of moisture inside the cable will actually change the K (dielectric constant) factor of the cable and, thus, the capacitance will change again without any change in L. Once again, this will change the impedance. It should be apparent that it is quite easy to change the C of the cable, but very difficult to change its L. Therefore, we have to be very careful in maintaining the relationship between L and C so as not to upset our 75 ohm impedance. This becomes important in our consideration of the cable, because every time a mismatch in the cable is produced, some of the power which would normally be expected to continue down the line does not. Instead, some signal power returns from the point of mismatch in the direction from which it came. We identify the returning signal at the point of mismatch by the term *return loss*. As an example, if the return loss from the mismatch were 20 dB, this could indicate that the signal being reversed and returning in the direction from which it came, would be 20 dB below the signal that arrived at the point of mismatch. Obviously, the amount of signal available to continue down the line must be reduced by the amount of the signal diverted. The presence of mismatches literally steals signal power that has already been paid for. It might actually be necessary to add amplifiers to the cascade in order to make up for this accumulation of mismatch losses. This is a minor problem compared to that which arises because of the presence of two mismatches in a single length of cable. Signal 1, arriving at mismatch B, has a portion of its power turned around and headed in the opposite direction. The return signal, arriving at mismatch A, now comes back down the transmission line as signal number 2. At mismatch B, then, signals 1 and 2 are the same in content, but different in *amplitude* and *time*. Both factors are important.

First let us consider the amplitude change. Assume that mismatch B is quite bad and that the return loss at mismatch B is equal to 10 dB. The return signal, then, starts out in its way from mismatch B to mismatch A, 10 dB below signal 1. It must, of course, go through the cable, or transmission line. Then, let us assume that the cable loss is 5 dB at the frequency of interest. When we arrive at mismatch A, the signal will be 15 dB below signal 1, which arrived at mismatch B. However, at mismatch A, we also have a bad mismatch and the return loss at there is 10 dB. This means that signal 2, starting out from mismatch A, is reduced from signal 1 arriving at mismatch B by a total of 25 dB; i.e., the combination of the return loss at mismatch B, the cable loss returning to mismatch A, and the return loss at mismatch A. That signal must return through the transmission line, so it is once again subjected to a cable loss of 5 dB. Hence, the total difference between signal 1 at mismatch B and signal 2 at mismatch B is 30 dB. This would be considered satisfactory in any feeder line, but would not be considered satisfactory in a transmission line or the main trunk line system.

We must next concern ourselves with the time difference between signal 1 and signal 2. The speed with which signals travel in free space is approximately 1000 feet per microsecond (also *c*, the speed of light). To find out how fast signals travel in the transmission line, we simply refer to the velocity of propagation factor for this particular cable. Most of the cables used to present employ a foam dielectric and have a propagation velocity of .82. This means that signals travel in the transmission line at a speed of 82% of what they would in free space ($c/.82$), or 820 feet per microsecond. This particular piece of cable might be something on the order of 400' long. In order to go from mismatch B to mismatch A and return, the signal must travel 800'. So, the difference in time between signals 1 and 2 might very well be about 1 microsecond.

To understand what happens next as a result of this difference in time, we must learn something about how a TV picture is made.

The TV camera has a photo-sensitive as part of its element, which literally displays an image of the picture that it is seeing. It then electrically scans the picture by a small scanning dot, which moves from the extreme left-hand side of the picture to the right-hand side within a given time element. The time necessary to go from one side of the picture to the other is 52.39 microseconds. As it moves along this path, the light and dark areas of the picture are transformed into high or low peaks of voltage. When the scanning dot reaches the extreme right-hand side, it blanks itself out and returns to the left-hand side of the screen in 11.11 microseconds. The complete horizontal time interval, from the time the scanning dot leaves the left-hand side until it returns again, is a total of 63.5 microseconds. In addition to this information, the TV camera transmits a horizontal and vertical synchronized pulse. Its purpose is to keep the TV receiver in step with what the camera is seeing. As a result, a scanning dot is reproducing the picture that the camera sees on the TV receiver, and the synchronized pulse is precisely following the scanning dot in the camera. In other words, when the scanning dot of the camera starts out on the left, it also starts out on the left-hand side of your TV receiver. When it arrives at the right-hand side, then returns to the left-hand side once again, it does the same thing in the TV receiver. The electrical impulses which the camera has put out are imposed on the TV receiver screen to literally print the picture that the camera has seen. If, however, the same information arrives at the TV receiver twice, as it does in the case of the reflected signal, then the set sees the same signal on both occasions, but the two arrive at different times. Therefore, a second picture exactly like the first appears, but it is slightly displaced in time on the TV screen. This is what we call "ghosting." Because the second signal was delayed by 1 microsecond in the example, we can now determine how far in time the second signal will be displaced. In terms of locus, the delayed signal is actually displaced by approximately 5/16. This information can be used in a different fashion to explain something about the system being worked on, depending upon the level of delay seen. For example, the longest space possible in a cable system, where discontinuities appear, is between two amplifiers. Because amplifiers are one-way devices, the reflection must be from the input of one amplifier back to the output of the previous amplifier, and then return. Let us assume that the cable we are using would have about 1800' spacing, and that it is ¾ aluminum with a velocity propagation factor of 82%. The longest delay possible, would be 4.4 microseconds. In terms of a 21 TV screen, because the delay is approximately 8% of the total width of the screen, the maximum ghosting possible on a system using this kind of cable would be one which was displaced from the original image by 1 ⅜. Anything greater could not possibly have been created on the cable system itself, but must have originated prior to the input of signal at the head-end. Even 1 ⅜ on the cable would be highly unusual, as in order to get this kind of displacement, we would have to have amplifiers separated by pure cable. The cable itself would have considerable attenuation at the frequencies normally of interest. For example, on channel 13, the attenuation through the cable twice — which is what is required for a ghost — would be 40 dB; and this, added to whatever the input and output impedance matches of the amplifiers were, would make a visible ghost impossible. However, at channel 2, the cable would have about 10 dB of loss. Therefore, the round trip would give us 20 dB of isolation between the amplifiers. Thus, if we had a bad input and output match in a pair of amplifiers, we could likely be in difficulty. This emphasizes the need for good impedance matching at the amplifiers themselves. If we have a nominal impedance match of 16 dB at each amplifier, we would then find that the total attenuation to the reflected signal would equal the 20 dB of the cable plus the 16 dB of output match at the preceding amplifier for a total of 52 dB. This, of course, would be adequate to provide us with a system free of ghosting. However, anything done on cable — such as splicing or splitting with directional couplers or splitters, or introducing any piece of equipment into that span of cable, which reduces the total length of cable from one device to the other — has a deleterious effect. Any mishandling of such connections can create a situation where the return path (back and forth through the cable), is reduced considerably from the 20 dB normally expected between two amplifiers. Therefore, the ghost is introduced once again. In any event, the displacement of such a ghost must necessarily be something less than 1 ⅜ on the screen.

This information is very useful to the CATV system technician, because knowing that any reflected signal must go back and forth through the cable twice, he or she can then insert a pad between any two suspected discontinuities. The reflected signal must go through that pad twice, and this automatically improves the signal to reflection discontinuity by twice the value of the pad. Using this technique, the insertion of a pad, which will improve the picture quality, has located the proper discontinuities. An improvement in picture quality does not in itself indicate that the discontinuity creating the ghosts has yet been found and, therefore, you can ignore everything on the line in which you have inserted that pad.

19.21 Standing Waves

Consider what happens when the terminating resistance R is not equal to Z_o. The load R no longer appears to the section of line immediately adjacent to it as simply more line. Such a line is said to be *mismatched*. The more R differs from Z_o, the greater the mismatch. The power reaching R is not totally absorbed, as it was when R was equal to Z_o, because R requires a different voltage to current ratio than when power is traveling along the line. The result is that R absorbs only part of the power reaching it. The remainder of the reflected power acts as if it had bounced off a wall, starting back along the line toward its source. The greater the mismatch, the larger the percentage of incident power reflected. In extreme cases, when R is zero, a short circuit, or infinity (an open circuit), all of the power reaching the end of the line is reflected.

19.22 Reactance

Reactance can be defined most simply as the frequency dependent tendency of a circuit to retard the flow of current.

19.23 Resistance, Reactance, and Impedance

To understand reactance, let us look at *resistance*, and *impedance*. Resistance, as stated by Ohm's law, is the characteristic of a circuit that limits the flow of current. Resistance may be the given value of a circuit component (the *resistor*), or the quantity of resistance in wire, cable, or other circuitry. Resistance by definition gives us a value in ohms Ω at zero frequency (dc).

When dealing with frequencies greater than (ac), resistance is no longer sufficient to give a complete picture. Hence, the term "reactance" must be employed. The combination of resistance and reactance is impedance (Z). At zero frequency (dc), impedance and resistance are the same, but at frequencies greater than zero (ac), reactance comes into play, so that impedance (Z) equals resistance (R) plus reactance (X).

19.24 Inductive Reactance

Inductive reactance (X_l) is that portion of total circuit reactance which is provided by coils, chokes, and transformer windings. Any device in which wire is circularly wound around some type of core is an *inductor*. Inductors are current devices and inductive reactance (X_l) increases as frequency increases. Thus, an inductor passes low frequencies more easily than high frequencies. The reason for this is the fact that current flowing through the wire generates a magnetic field about the inductor; the intensity of the magnetic field is determined by the value of the current and the direction of the field is determined by direction of current flow. For a constant (dc, zero frequency) value of current, the intensity and direction of the magnetic field are constant. Consequently, current flows easily through the inductor because the inductor appears to the current to be nothing more than a length of wire. Magnetic fields by nature resist change, and when they do change, they do so slowly. Thus, in order for an inductor to conduct a changing (alternating) current, the magnetic field must change as the magnitude and direction of current change. The magnetic field can change more easily at low (slow rate of change) frequencies than at high (fast rate of change) frequencies. Therefore, the inductive reactance X_L (in ohms) of a circuit, increases with both frequency and the value of the inductor (in henries).

19.25 Capacitive Reactance

Capacitive reactance (X_c) is that portion of total circuit reactance caused by capacitance. Any time that two conducting surfaces are placed parallel to each other, and are separated some small distance by a non-conducting substance, the result is *capacitance* (exemplified by the circuit component known as a *capacitor*). Capacitors are voltage devices. Capacitive reactance (as opposed to inductive reactance) decreases as frequency increases. Therefore, a capacitor tends to pass high frequencies more easily than low frequencies. Remember, a capacitor is simply two parallel conductors separated by a non-conductor. Therefore, it is basically an open circuit, and current cannot flow through it. At dc (zero frequency) a capacitor simply acts as an open circuit and consequently blocks a dc current or voltage level. When voltage is applied to a capacitor, an electric field is generated between the two conducting surfaces (this is what occurs when a capacitor "charges"). Electric fields, like magnetic fields, resist change in intensity, but will do so slowly. Thus, if a changing voltage (ac, frequency greater than zero) is applied to one side of a capacitor, the electric field in the capacitor resists change, and in order to maintain a constant intensity between the two conducting surfaces, pulls the other side of the capacitor along with the first side. Thus, a changing voltage applied to one side of a capacitor appears on the other side and causes a corresponding current to flow in the far side of the circuit. (Be aware that no current flows through capacitor.) Because the electric field has more time to change at low frequencies, the output of a capacitor follows the input better when the frequency is high. Thus, as frequency increases, X_c decreases. The actual value of X_c (in ohms) is inversely proportional to the value of capacitance (in farads) as well as frequency.

19.26 Resonance

Impedance equals *resistance* plus *reactance*, and *reactance* is a combination of *inductive reactance* and *capacitive reactance*. For a given capacitor and a given inductor there is some frequency where $X_L = X_c$. This frequency is known as the "resonant frequency," and this concept termed *resonance*.

If the capacitance and inductance are in series, at resonance $X_L = X_C$, X_L and X_C cancel and the impedance is purely resistive (i.e., minimum impedance). The resonance characteristic can be used to design a network which passes only a certain band of frequencies, while blocking all frequencies outside the band. The network can be designed such that the band is either narrow or wide. The center frequency of this band is the *resonant frequency*.

On the other hand, if the capacitance and inductance are in parallel, at resonance $X_L = X_C$ once again, but this time X_L and X_C are added together causing the impedance to reach a maximum. This characteristic can then be used to design a network which blocks a certain band of frequencies, while passing everything outside the band. In the parallel case, as with the series case, the band may be either narrow or wide, and the center frequency of the band is the resonant frequency.

An important application of such networks is to separate the audio and video portions of TV signals.

19.27 Network Q Factor

The above discussion of resonance mentioned networks that can be used to either pass or block a specific band of frequencies, containing desired signals which we want to pass, and very close outside this band are signals which we want to block. Then, clearly, the network used must be very selective. In other words, the response of the network must be very sharp so that it does not partially pass unwanted signals and partially attenuate the desired signals. The amount of sharpness, or selectivity, of this network is measured by a factor called Q, i.e., "quality." Q is sometimes known as the *figure of merit* of a network; the bigger the number, the greater the selectivity. Thus, in designing these networks, also known as *tuned circuits*, we usually strive for a high Q factor.

20 Antennas

20.1 Antenna Fundamentals

The major use of antennas in the CATV system is to receive broadcasts off the air. The receiving antenna is simply a conductor that receives RF signals in the form of waves. These waves induce a current flow in the conductor, which varies according to the modulation of the signals received. The variations represent intelligence (or information), which can eventually be reproduced on a subscriber's television set, as a replica of the intelligence (or information) which was broadcast.

20.2 Types of Receiving Antennas

Many different types of antennas are used in CATV systems. The particular type should be chosen to meet the individual reception requirements for a specific channel or group of channels at a given location. The basic antenna used in the earliest systems was the *Yagi*, which provided fairly high gain in a relatively small package. The basis of the Yagi is a single *dipole*.

20.3 The Dipole Antenna

Technicians have all heard the phrase "cut to channel." The frequency that a dipole is used to receive is determined by the size (length) of the dipole. This size is inversely proportional to the frequency; i.e., as the frequency increases, the dipole size decreases. The dipole is a bidirectional antenna with equal efficiency in the direction perpendicular to its own axis. There are two equal and symmetrical lobes on either side of the dipole. The dipole has good reception on an axis perpendicular to its own and very poor reception on an axis parallel to its own. The action of the dipole can be improved by the inclusion of a second element called a *reflector*.

20.4 The Reflector

The reflector is located on the side of the dipole opposite the direction from which the signals will arrive. Its length will be greater than that of the dipole itself. This gives it the electrical characteristic of appearing to be a short circuit to signals arriving from the direction in which the dipole is placed.

20.5 Reflector Action

Any signal missing the dipole and continuing on through, or any signal re-radiated by the dipole, will arrive at the reflector one-quarter of a wavelength after having passed the dipole itself. Because the reflector has the electrical characteristic of a short circuit, it will take such signals and invert their phase, which is the equivalent of one-half wavelength. Therefore, signals re-radiated from the reflector will be changed in phase from the signals at the dipole by a quarter of a wavelength as they return to the reflector, a half wavelength by the inversion created by the reflector, and by an additional quarter wavelength in returning to the dipole. The total change is one full wavelength, so that signals re-radiated by the reflector returning to the dipole are in phase with the signals arriving at the dipole. Thus, they reinforce those signals and provide additional gain. This has the effect of taking the second pole pattern, or lobe, and doubling it so that it adds to the pole pattern, or lobe, that exists in the forward direction. It serves no useful purpose to add further reflectors to a Yagi antenna. Their action (because of the distance involved in the phase change) is such that they would not only fail to help the gain of the antenna, but would actually reduce it. Hence, every Yagi antenna has a single reflector element.

20.6 Radiators

To further improve the action of the Yagi, elements called *radiators* are added. The radiators are shorter in length than the dipole itself, which provides the electrical characteristic of an open circuit, and they are spaced to give optimum gain. This spacing generally turns out to be about one-eighth of a wavelength or less.

20.7 Radiator Action

The radiators intercept signals and re-radiate them in phase with the received signal. This signal then travels towards the dipole, arriving in phase with the signals which the dipole itself would have intercepted, thereby reinforcing the signal the dipole is seeing and increasing the gain of the array. Each of the radiators then performs the same function; i.e., it reinforces and increases the signal at each radiator. Hence, the signal is constantly reinforced, proceeding from radiator to radiator, until we return to the dipole itself. All signals from the radiators act to reinforce the signal at the dipole. This is a simple description of how the Yagi antenna obtains its gain.

However, as we proceed further away from the dipole, and with more and more radiators, we find that the addition of more radiators will, at some point, arrive at the point where there is no signal return. Hence, the number of radiators is limited by the point of no return. As a rule most Yagi antennas will have something in the neighborhood of 5 to 6 radiators on low band antennas and 10 to 12 radiators on high band antennas, simply because going beyond that point provides very little additional gain at the expense of high mechanical instability.

20.8 The Yagi

Therefore, the Yagi is an excellent antenna from both the standpoints of directivity and gain. It has excellent gain characteristics in a direction which is perpendicular to the elements of the antenna. It has poor pick-up qualities from behind, and even poorer pick-up from the point parallel to its own axis. A typical polar pattern might appear to be one with a very nice front lobe and then several smaller lobes on its back side.

20.9 Stacking

The directivity of the Yagi can be improved by using extra antennas in a stacked array. This stacked array can be spaced to favor either additional directivity or additional gain.

If antennas were merely spaced for gain, we would place two antennas in the vertical plane, such that the spacing between the antennas was approximately one-half wavelength. This would give optimum gain of approximately 3 dB. If spaced horizontally, the two antennas can accomplish the same thing by spacing them at approximately nine-tenths of a wavelength wide for optimum gain. Once again, a gain of approximately 3 dB can be realized.

However, in many cases it may be wise to ignore this optimum gain spacing and instead use the antennas for directivity in order to minimize unwanted signals. This is done very simply in the spacing of the antennas themselves.

Vertical spacing generally is not used to eliminate signal interference, simply because there is usually insufficient difference between the bearing of the desired signal and the bearing of the undesired signal to provide for any rejection in the antenna system itself, except in very rare cases. However, when two antennas are stacked vertically, the lobes sharpen and become longer. This automatically ensures efficient reception from the target area and less reception from the unwanted area. When we space horizontally, even more good is accomplished because now the sharpening of the lobes has the effect of giving us more gain on the desired channel and less gain on the undesired, thereby improving the desired-to-undesired ratio. Therefore, the actual gain may not be optimum, but the overall performance of the stacked array is best for directivity.

20.10 Log-Periodic Antennas

The Yagi is certainly less expensive than the log-periodic antenna. Nevertheless, there are so many advantages to the system operator in using the log-periodic type of antenna that the additional cost is more than justified. (See further discussion in Ch. 23.)

20.11 Co-Channel Rejection

The log-periodic response of the individual antennas, together with the particular arrangement of the elements within the array (the diamond-four configuration), provide for lower sidelobes, reduced wide-angle radiation, and better front-to-back ratio. All these factors ensure excellent co-channel rejection.

20.12 Bandwidth

The log-periodic antenna is essentially a broadband antenna, as compared to a Yagi which is a narrow-band antenna. Where an individual Yagi, or array of Yagis, is needed for each channel received by a system, only two log-periodics are necessary for full coverage of the low VHF band and only one for the entire high VHF band.

20.13 Impedance Match

Impedance match makes the log periodic much better for color reception than the Yagi. Yagis are normally 300 ohms and a transformer is required in order to use 75 ohm cable. Log-periodics, on the other hand, have a 75 ohm output impedance.

20.14 Gain

An array of log-periodics will have slightly less (1 or 2 dB) gain than an array of Yagis; but, as we have pointed out in our discussion of Yagis, directivity can be more important than gain.

20.15 Icing

Log-periodic antennas are virtually unaffected by icing because of the low Q of the transmission line. Yagis, however, may experience a complete phase shift under severe icing conditions. This may be so severe as to cause the sound signal to be lost.

20.16 Mounting

The log-periodic antenna, because it is fed from one end, can be mounted in a cantilever fashion; i.e., the antenna is attached to the mast at its "cold" point and projects outward from the mast. The Yagi, because it is a center-feed device, must be mounted on a boom which projects out from the mast and requires guy wires for stability. This type of mounting also may impair the antenna's directivity.

20.17 Search Antenna

The name *search antenna* has been used to describe a rotatable antenna, located on the very top of a tower. The search antenna is usually an all-band antenna, covering both the low and high VHF channels as well as the UHF channels. We prefer the terms "reference antenna" or "standby antenna," instead of the term "search antenna." The basic use of this standby or reference antenna is to provide a reference with which the performance of other individual antennas can be compared.

Consider a situation where channel 2 on the system is suddenly obscured with co-channel interference. The question is, did something happen to the array or is this merely a propagational freak condition? By observing the picture quality of channel 2 on the reference antenna, you can determine whether the array is in trouble or if this is merely a propagational freak. If channel 2, as received off the reference antenna, has less co-channel interference than the array, then there is little question that something has happened in the stacking harness of the channel 2 array. The search antenna can also be moved toward the direction of the probable source of co-channel interference and determine whether the suspected offending channel is stronger than usual.

Consider a second situation where channel 10 is much weaker and snowier than normal. Has something happened to the station? Has something happened to channel 10's antenna? Is it just due to unusual propagational conditions? By observing channel 10's picture quality from the search antenna, a good indication can probably be acquired as to the particular trouble.

If the picture quality from the reference antenna is actually better than the picture quality from channel 10's antenna, then, obviously, either the channel 10 antenna, its down-lead, or its preamplifier is the source of trouble. If the picture quality from the reference antenna is worse than that received from channel 10's antenna, either something has happened to the propagational conditions, or perhaps the transmitter power has been reduced.

To take advantage of all of the possibilities that the reference antenna offers, you must routinely compare the performance of the reference antenna on each channel and compare each with the performance of its individual channel counterpart. It is a good idea to note these readings once or twice every month in a log book. This book should be readily available in times of reception difficulty.

20.18 Beamwidth

When the radiated power of an antenna is concentrated in a single major lobe, the angular width of the lobe is called the *beamwidth*, a term which is applicable only to antennas whose patterns are of this general type. Some antennas have a pattern consisting of many lobes, all of which are more or less comparable in their maximum power density or gain and not necessarily of the same angular width.

If an antenna has a narrow beam and is devised for reception, it can be used to determine the direction from which the received signal is arriving; i.e., will provide information about the direction of the transmitter. In some applications, a receiver may be unable to discriminate completely against an unwanted signal that is being transmitted at or nearly the same frequency as the desired signal. In such a case, pointing a narrow receiving antenna beam in the direction of the desired signal is helpful, because the resulting greater gain of the antenna for the desired signal and the reduced gain for the undesired signal may provide the necessary discrimination. If the directions of the desired and undesired signals are widely separated, even a relatively wide beam will suffice. The closer the two signals are in direction, the narrower the beam must be to provide effective discrimination.

There are situations that require the use of wide beams. For example, a broadcasting station must radiate a signal simultaneously to listeners in many different directions — typically, over a 360° azimuth sector. Any narrowing of the beam to obtain gain must be done in the vertical plane. At the low frequencies of the regular broadcast band (approximately 550 to 1600 kHz), and lower, such beam narrowing in the vertical plane is not feasible, because it requires an impractically large, or highly placed, antenna. In TV and FM broadcasting in the VHF and UHF bands, this means of obtaining gain while maintaining 360° azimuthal coverage is often used.

An antenna beam is typically "round-nosed." Thus, defining beamwidth is a problem. In this case, it would be possible to cite the width of the beam between the two null points on either side of the maximum, because these are two definitely measurable points. Not all beams have such null points, however, although they are ordinarily present. It is logical to define the width of a beam in such a way that it indicates the angular range within which radiation of useful strength is obtained, or from which good reception may be expected.

From this point of view, a conventional method has been adopted for measuring beamwidth between the points on the beam pattern at which the power density is half the value of the maximum. On a plot of the electrical intensity pattern, the corresponding points are those at which the intensity is equal to one-half, or 0.707 of the maximum value.

The angular width of the beam between these points is called the *half-power beamwidth*. When a beam pattern is plotted with the ordinate scale in decibels, as is frequently done, the half-power points correspond to the –3 dB points. Because of this, the half-power beamwidth is often referred to as the "3 dB bandwidth."

20.19 Minor Lobes

As mentioned in the discussion of antenna patterns, a directional antenna usually has, in addition to a main beam or major lobe of radiation, several smaller lobes in other directions. These are the minor lobes of the pattern. This lobes adjacent to the main lobe are called *sidelobes*, and those that occupy the hemisphere in the direction opposite to the main beam direction are called *back lobes*. Minor lobes ordinarily represent radiation, or reception in undesired directions. An antenna designer should attempt to minimize minor lobes; i.e., to reduce their level relative to that of the main beam. This level is defined in terms of the ratio of the power densities in the main beam maximum and in the strongest minor lobe.

Because the sidelobes are usually the largest of the minor lobes, the ratio is often called the *sidelobe ratio*. A typical sidelobe level, as found in an antenna where some attempt had been made to reduce the sidelobe level, is 20 dB. This means that the power density in the strongest sidelobe is 1% of the power density in the main beam. Sidelobe levels of practical, well-designed directional antennas typically range from 13 dB (a power density ratio of 20) to 40 dB (a power density ratio of 10,000). A sidelobe level better than 30 dB requires very careful design and construction. In some applications sidelobes are not necessarily harmful unless their levels become comparable to the main beam level. In other applications, the sidelobe level may need to be held to an absolute minimum.

20.20 Gain and Directivity

An *isotrope* is an antenna which radiates uniformly in all directions. An isotropic pattern is a perfect spherical surface; i.e., if the electrical intensity of the field radiated by an isotrope is measured at all points on an imaginary spherical surface with the isotrope at the center, the same value will be measured everywhere.

Such a radiator cannot be physically constructed — it is a theoretical entity. All real, as opposed to theoretical, antennas have some degree of non-uniformity in their three-dimensional radiation patterns. Because it is possible for an antenna to radiate uniformly in all directions in a plane, it is possible to design an antenna that has approximate omnidirectionality in three dimensions; but perfect omnidirectionality in three-dimensional space can never be achieved. The concept of such an ideal omni-directional radiator, an isotrope, is most useful for theoretical purposes.

A *non-isotropic antenna* will radiate more power in some directions than in others and has a directional pattern. Any directional antenna will radiate more power in its direction of maximum radiation than an isotropic antenna, if both antennas are radiating the same total power. Because the directional antenna sends less power in some directions than the isotropic antenna, the directional antenna must send more power in other directions if the total power radiated from both antennas is to be the same.

This conclusion can be further demonstrated by the following example. If an isotrope radiates a total power P_i watts and is located at the center of an imaginary sphere whose radius is R meters, the power density over the spherical surface is *power of isotrope* = $P_i/4R^2$ watts per square meter, because the total P_i is distributed uniformly over the surface area of the sphere, which is $4R^2$ meters.

Imagine that it is possible to design an antenna that radiates the same total power uniformly through one-half of the spherical surface with no power radiated into the other half. This ideal radiator might be called a semi-isotrope. Because the half-sphere has a surface area of $2R^2$ square meters, the power density would be *power of semi-isotrope* = $P_i/2R^2$ watts per square meter.

21   Regions and Propagation

The region immediately next to the transmitting antenna is known as the *line-of-sight* region. This region extends out to the *radio horizon*. The distance from the transmitting antenna to the radio horizon is given by the formula:

$$D = \sqrt{2h_T} \tag{21-1}$$

where D is the distance to horizon in miles, and h_T is the height of the transmitting antenna in feet. The radio horizon is assumed to be on the ground if the earth is a relatively smooth sphere. Any obstruction along a given path must be taken into consideration.

If the receiving antenna is also elevated, the maximum radio line-of-sight is given by the formula:

$$D = \sqrt{2h_T} = \sqrt{2h_R} \tag{21-2}$$

where h_R is the height of the receiving antenna in feet.

Waves emitted by television transmitters exhibit this line-of-sight action, as shown in Fig. 21.1. However, the cut-off at the horizon is not sharply defined. In traveling through the air, the wave is bent slightly toward the earth, and the VHF range is considered to extend approximately 15% beyond the line of sight or optical horizon. This extended range is the radio horizon.

If the propagation path in the line-of-sight region is sufficiently free from objects that might absorb or reflect radio energy, this region can be further subdivided into a region known as *free space*. The free space region is seldom realized because of the presence of the earth's surface. The usual condition for frequencies above approximately 30 MHz is such that one wave travels

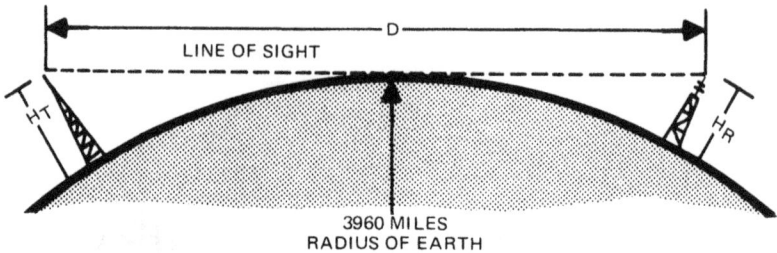

Figure 21.1 Line of Sight
Source: National Cable Television Institute

directly from the transmitter to the receiver, and a second wave from the same transmitting antenna strikes the ground between the two antennas and then is reflected to the receiving antenna. In this case, the ground acts partially as reflector and partially as absorber. The ground-reflected wave has traveled farther and, therefore, the phases of the two waves at the receiving antenna are different.

The result of this phase difference is an oscillating signal level whose amplitude and frequency vary according to the height and distance from the transmitting antenna. As the energies of these two paths unite in phase, the result is a maximum phase difference; as they unite out of phase, the result is a minimum phase difference.

21.1 Field Strength

When the distance between the transmitting and receiving antennas is considerably greater than the antenna heights, the angle of incidence to the surface of the reflected wave will be small. The reflection at the earth can then be assumed to take place with a phase reversal and no change in magnitude. Under these circumstances, the two waves at the receiving point have equal amplitude, but generally differ in phase. The difference in path length for the direct and reflected waves is less than 1/6 wavelength, and the field strength for an ideal flat earth becomes inversely proportional to the square of the distance, and so drops off relatively rapidly.

The space beyond the line-of-sight region is known as the region *beyond-the-horizon*. This region is shadowed from direct rays by the curvature of the earth or other obstructions.

The *diffraction region* lies adjacent to and below the radio horizon. This is the region in which most CATV towers are constructed. Energy reaching this region must be bent or reflected by some process. One such process is *diffraction*, which is a fundamental property of wave motion. Sharp shadows, such as would be created by a beam of light striking a solid object, are not created when RF waves encounter large obstructions. Reception behind the obstruction is possible for a short distance; but there is a shadow which is somewhat fuzzy and a gradual, rather than sharp, transition in signal level.

21.2 Scatter Propagation

Diffraction was considered the only mechanism whereby VHF and UHF energy can be supplied beyond the horizon. One type of diffraction is *tropospheric scatter*. Tropospheric scatter is caused by random irregularities of the dielectric constant of the atmosphere.

It should be noted that two types of signal fading are encountered in scatter propagation. The first is the rapid fade caused by multi-path transmission in the atmosphere. The multi-path condition is caused when a signal is reflected from many signals which are random in location and have random motions. The received signal is the sum of these random reflected signals. The received signal may change from maximum to minimum and back to maximum in a matter of seconds. Fast multi-path fading tends to reduce the allowable bandwidth to less than 5 MHz. This reduction in bandwidth is caused by the time delay associated with the different paths.

The second type of fading encountered in scatter propagation is slower, spreading over a period of hours or even days. These slow changes in signal level result from a combination of variations in atmospheric refraction from day to night, and of humidity and temperature changes along the scatter path.

There is one other important mechanism by which waves are propagated beyond-the-horizon. Waves can be propagated by reflections from a portion of the ionosphere known as the *Sporadic-E layer*. Clouds of very high ionization are quite well known, but their cause is still subject to speculation. The E-layer of the ionosphere is located 50 to 70 miles above the surface of the earth. Bombardment of this region by radiation from the sun produces ionized molecules. Sometimes clouds of very high ionization are sufficiently dense to reflect signals in the 50 to 100 MHz range. Sporadic-E hop (or skip) distances vary, from a minimum of 500 miles to a maximum of about 1400 miles for a single hop.

The portion of the earth's atmosphere extending from sea level to a height of about six miles is the *troposphere*, or weather layer. This is a region of wind, storm, and rain. The temperature of the troposphere decreases about 20°F per mile of increasing altitude and reaches a minimum value near 58°F at the upper limit of the region.

Directly above the troposphere is the *stratosphere*, or constant-temperature

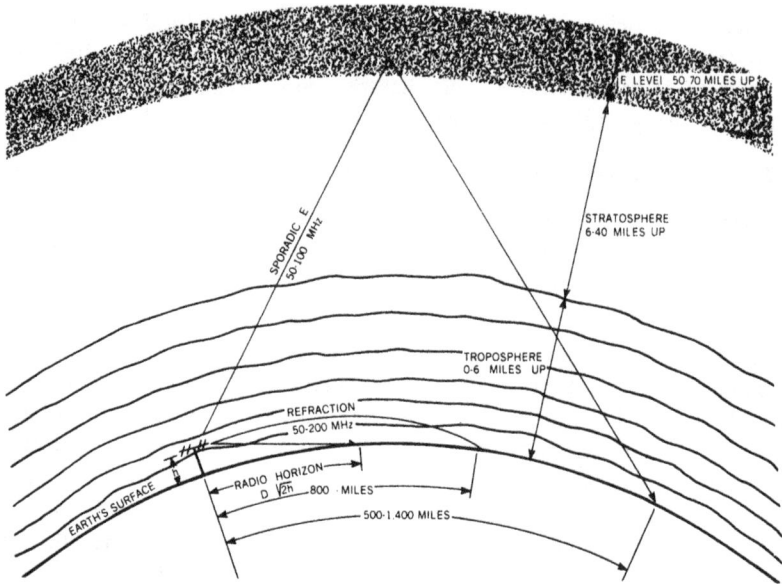

Figure 21.2 Scatter Propagation in the Earth's Atmosphere
Source: National Cable Television Institute

zone. The stratosphere extends to an approximate height of 40 miles. This region has little effect upon VHF propagation.

The E-layer of ionosphere is located 50 to 70 miles above the surface of the earth. Bombardment of this region by radiation from the sun produces ionized molecules. Sometimes clouds of very high ionization are sufficiently dense to reflect signals in the 50-100 MHz range.

In free space, or a vacuum, a wave expands outwardly from its source. Each part of the wave travels along a radial line and has a constant velocity equal to that of light.

In the ionized gas of the ionosphere, a wave will travel as it does in free space, but with reduced velocity. The ratio of the free space velocity to the velocity in the dielectric medium is called the *index of refraction* of the medium.

In a windless standard atmosphere, the temperature and water vapor content decrease steadily with increasing altitude. This decrease in temperature and water vapor, associated with altitude increases, causes a decrease in the index of refraction with altitude. This results in the velocity of transmission increasing with height above the ground, and as a result the wave is bent or refracted toward the earth. As long as the change in dielectric constant is linear with

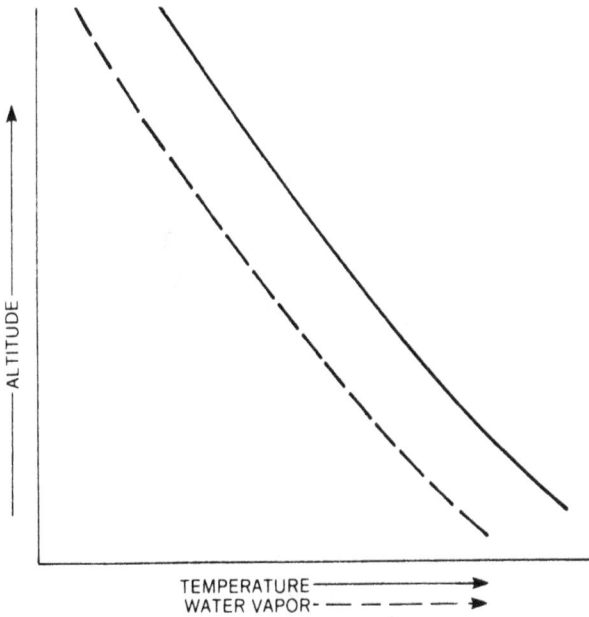

Figure 21.3 Temperature and Water Vapor vs.
Source: Scientific Atlanta

height, the net effect of refraction is the same as if the wave continued to travel in a straight line, but over an earth whose radius is 4/3 (1.33) times the true radius.

Most technicians have noticed that signal strengths are higher in the evening and early morning than during mid-afternoon. This phenomenon is caused by the following conditions: as the sun goes down, air immediately adjacent to the earth cools rapidly, while the air at higher altitudes cools much more slowly. This causes the dielectric constant of the air near the earth to increase, creating change in dielectric constant with altitude. With this increasing dielectric gradient, the effective earth radius increases. In the early morning hours, the sun warms the air at higher altitudes before the air adjacent to the earth is warmed. Consequently, the effective earth radius is again larger than 1.33 times the true radius.

When the dielectric constant decreases about 10^{-7} per foot of height — in the standard atmosphere the decrease is 10^{-8} per foot — it has the effect of making the earth flat. Under such a condition, a wave that starts parallel to the earth will remain parallel. When the dielectric constant decreases more rapidly when 10^{-7} per foot of height, radio waves that are radiated parallel to, or at an

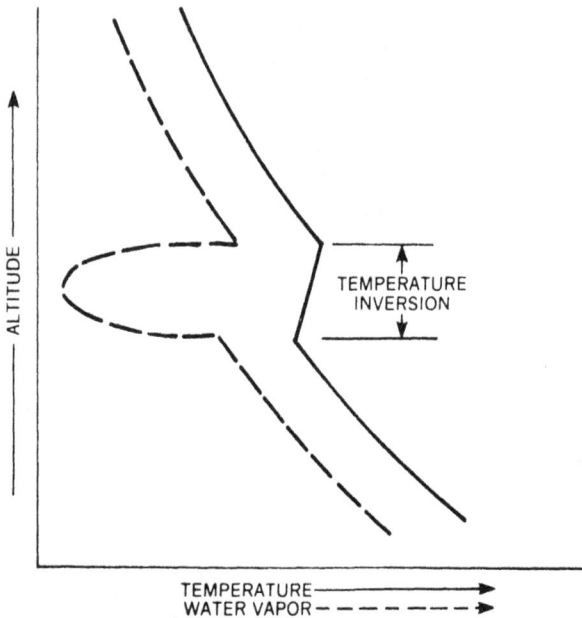

Figure 21.4 Temperature Inversions vs. Altitude
Source: Scientific Atlanta

angle above, the earth's surface, may be bent downward sufficiently to be reflected from the earth. This phenomenon is known as *frequency inversion.*

After reflection, the wave is again bent toward the earth as it passes through the atmosphere, and the resulting path of a typical wave is similar to the path of a bouncing tennis ball. The radio energy appears to be trapped in a duct, or waveguide, between the earth and the troposphere. This phenomenon is referred to as either *trapping* or *ducting.*

22 Head-End Testing and Alignment

22.1 The RF Spectrum of a Television Signal

The RF spectrum of a television signal, as transmitted in the standard format used by the United States and Canada, is displayed in Fig. 22.1. The signals consist of a visual (picture) carrier, an aural (sound) carrier, and a chroma (color) subcarrier. This basic arrangement of carriers and subscribers, however, carries no video or other information. Additional energy in the form of sidebands is needed for any information to be carried. This additional information also requires spectrum space, because the carriers, *per se*, actually carry nothing other than themselves, but they are needed for the demodulation process in the television receiver.

22.2 Improved Head-End Performance Using a Test Demodulator

In addition to its traditional applications, a precision demodulator can benefit the CATV engineer as a tool for improving system performance. In the past, demodulators have been used as an input to a microwave link, and have found use in demodulator-remodulator types of head-ends. Also, a demodulator can be of great value during modulator evaluation and set-up, in aligning antennas, and in evaluating in-service head-end equipment.

Before entering into a discussion of these applications, a brief review of some issues concerning demodulator characteristics is in order. Two primary questions are the choice of envelope *versus* synchronous detection and the most desirable shape of the IF passband response. The latter question is of concern primarily with respect to envelope detectors. Both of these questions are currently being debated among the broadcast, TV receiver and CATV industries. Thus a divergence of opinion may be noted.

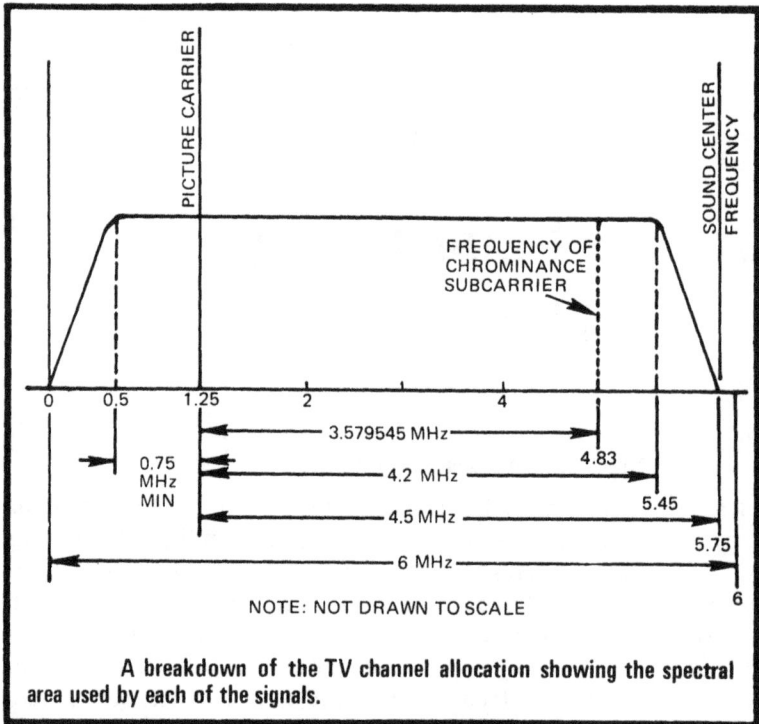

A breakdown of the TV channel allocation showing the spectral area used by each of the signals.

Figure 22.1 A Breakdown of the TV Channel Allocation Showing the Spectral Area used by each of the Signals

Source: Hewlett Packard, Cable Television System Measurement Handbook

The question of synchronous *versus* envelope detection is probably headed for a solution (though not permanent), with the conclusion that both types of detectors are necessary if a reasonably complete evaluation of the modulated signal is to be undertaken. For operational situations (e.g., a microwave relay link feed, either detector yields reasonable results, with the synchronous detector being preferred. Again, however, the presence of both detectors would be desirable.

22.3 Envelope Detection

Virtually all television sets in use (with the exception of a few marketed within the last two or three years) employ an *envelope detector*. This usually takes the form of a diode after the last stage of the IF amplifier, placed in series with the output. The IF signal is rectified by the diode, reproducing the envelope of the IF signal. Envelope detectors are characterized by simplicity, low cost, and relative immunity to incidental transmitter frequency modulation. Perhaps the most compelling reason for using an envelope detector in a test demodulator is the fact that it is so common in home receivers. This provides the opportunity to estimate the performance of the detector in a home receiver.

Nevertheless, envelope detectors have two significant drawbacks, one related to errors in the envelope of a *vestigial sideband signal*, and the other related to the practical problem of diode nonlinearity. The former problem is called *quadrature distortion*, and arises in any signal where the sidebands are filtered unsymetrically around the carrier.

Among the errors caused by quadrature distortion are ringing on transitions, differential gain (though this can also rise from diode nonlinearity), and chrominance-to-luminance crosstalk, which results in a darkening of those picture areas having considerable color.

One of the classical analyses of the quadrature distortion problem is to compute the actual *versus* ideal luminance and the red, green, and blue levels developed during envelope demodulation of color bars. This calculation shows considerable shift at all levels. However, the amount of shift is drastically reduced before detection. For this reason, the IF response of most television sets is reduced at the color subcarrier frequency.

The second problem associated with diode detectors is caused by the nonlinear forward response of real diodes. This generally takes the form of reduced diode efficiency at lower signal levels, resulting in additional differential gain and phase in the output as well as additional distortion of transient responses. These problems are visible near the white level, corresponding to minimum signal amplitude. The severity of the problem can be reduced through the use of better quality diodes operated at high signal levels.

22.4 Synchronous Detection

An alternative to envelope detection is *synchronous detection*. A phase-locked loop is normally used to reconstruct a CW reference signal having a definite phase relationship to the incoming picture carrier. This signal is mixed with the incoming carrier, and one resultant output is baseband video.

Everything said about the envelope detector applies in reverse to the synchronous detector. The synchronous detector, because of its cost and complexity, has seen virtually no application in consumer television sets, but has been available to the professional user since the 1970s. Recently, several integrated circuit manufacturers have introduced detector chips that claim to use synchronous detection. However, in these chips the reference signal is derived by introducing a sample of the modulated signal into an amplitude limiter. While this technique produced acceptable performance for consumer applications, experience to date indicates that these chips are inappropriate for precision applications, and do not yield true synchronous detection.

In addition to cost, complexity, and lack of consumer application, synchronous detectors degrade when the transmitter imposes incidental FM on its carrier. This is especially troublesome with some older transmitters, as a result of insufficient power supply filtering or insufficient power supply regulation. When incidental FM is present, both synchronous and envelope detector performance suffers, though the degradation is much greater in the synchronous detector case. This is because of the inability of the synchronous detector to follow the frequency modulation pattern. The severity of the problem is partially determined by the servo loop response of the phase-locked loop, compared with the effective wave shape of the frequency modulating signal.

On the other hand, the properly adjusted synchronous detector does not respond to quadrature modulation components. Thus, all the above problems mentioned related to quadrature distortion do not exist. Also, because the reference signal may be applied to the mixer at a very high level, and because the mixer may be designed to not operate diodes near their reverse breakdown voltage, the problem of diode nonlinearity is largely avoided.

22.5 Detector Choice

Thus, we face a choice of which detector to use. In general, we would prefer a synchronous detector for both off-the-air and test applications, because of its ability to demodulate pictures more accurately and the lack of possible masking of modulator faults.

On the other hand, the criterion by which technical performance must ultimately be judged is the picture seen by the consumer on his home TV set. Because most home TV sets use envelope detection, one could argue that an envelope detector would evaluate a modulator more similarly to how a TV set would evaluate it. Insofar as tradeoffs can be made in the transmitter delay predistortion to crudely compensate for envelope deficiencies, some transmitters might yield superior results when demodulated with an envelope detector.

The growing consensus in the broadcast industry is that a measurement demodulator must have both types of detection. This philosophy must surely follow for demodulators used by CATV systems for test applications as well. For off-the-air applications, the availability of both types of detectors is equally useful: the synchronous detector to use for highest quality demodulation during normal operation; the envelope detector as back-up to use if incidental FM is present.

22.6 Passband Response

Another question to be answered when specifying a demodulator, is the shape of the predetection (IF) *passband response*. As discussed above, improved response from an envelope detector is obtained when the IF response is reduced at the color carrier frequency. For this reason, most home TV receivers reduce the response at the color subcarrier. Unfortunately, no standards exist concerning the passband response, and receivers differ markedly in their response. The industry had been debating the passband response of a test envelope demodulator. In some manner, the response will probably be reduced at the color subcarrier, with post-detection restoration of the lost frequency response. Of course, as quadrature distortion does not affect synchronous detection, no similar problem exists with the synchronous detector. The IF response is held flat at the color subcarrier frequency, and a video amplifier with flat response is used.

22.7 Application

When modulators are used in a CATV system, a demodulator should also be available to aid in measuring modulator performance. The most fundamental measurement to be made on a continuing basis is *depth of modulation*. In a standard television signal, a maximum modulation depth of 87.5% is used. Modulating to greater depth may result in intercarrier *buzz* in the receiver, as well as causing *wash out* of the brighter areas of the picture, and excess differential gain and phase because of diode nonlinearity. Modulation to an insufficient depth will create a washed out or faded picture, loss of color brilliance, and possibly even loss of synchronization in the receiver. Some CATV modulators have no means for measuring depth of modulation. However, some more popular modulators have meters indicating *approximate* depth of modulation.

A meter of this type has its limitations. It may not detect a very brief excursion of some highlight into the over-modulated region, nor may it tell if the color carrier (only) makes excessive excursion. Also, the accuracy of a meter is only about 5%.

Many modulators incorporate a *peak white clipper* that does not permit over-modulation. However, the clipper can only be set while looking at demodulated wave form or while looking at the modulated IF with a high frequency oscilloscope. This latter technique must use the IF signal before the vestigial sideband filter, because the filter will effectively reduce the amplitude of the color subcarrier.

In order to use a demodulator for measuring depth of modulation, two references must be established. One reference is at the maximum amplitude of the RF carrier, corresponding to the synchronization tips. The other reference should be at a *known* depth of modulation, preferably corresponding to the cut-off (100% modulation) frequency. This reference is not available in the modulated signal, so it must be added by the demodulator. This is done with a *zone chopper*, a switch that interrupts the IF (or RF) signal momentarily, without affecting the AGC voltage. This interruption, or *zero chop pulse*, then establishes a reference.

Figures 22.2 and 22.3 show a demodulated signal featuring a zero chop pulse occuring during the vertical interval: Fig. 22.2 shows the vertical interval expanded; Fig. 22.3 shows the same signal with the wave form monitor set for a two-line display. The wave form monitor displayed in Fig. 22.2 has been lowered to show the zero chop pulse more clearly. If a wave form consisting of 40 units of synchronization and 100 units of video is modulated onto a carrier to a depth of 87.5%, then the zero chop pulse corresponds to 160 units of signal. Thus, by setting the synchronization and chops spaced by 160 units, the synchronization should occupy 40 units, the video 100 units, and the zero chop pulse should be 20 units above the maximum video. A synchronous detector is definitely preferred for making this measurement, because diode nonlinearity in an envelope detector will shift the zero chop level.

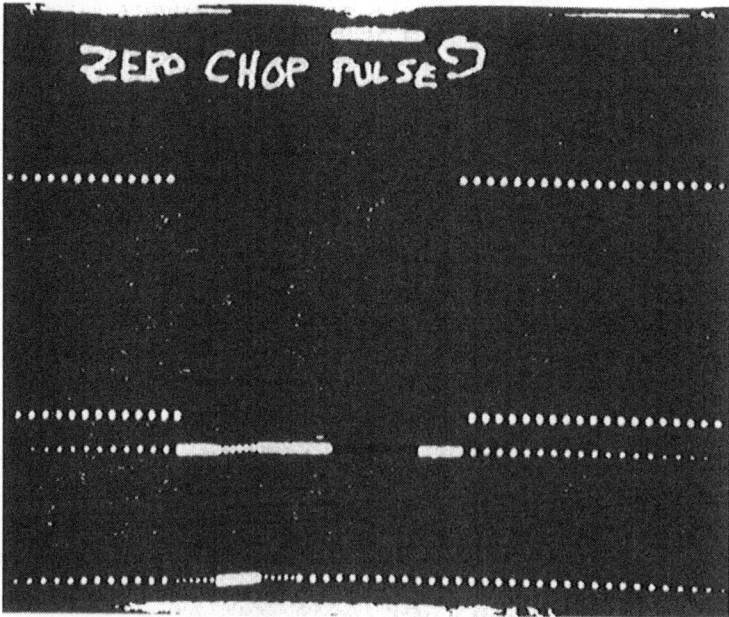

Figure 22.2 Demodulated Signal Featuring a Zero Chop Pulse Occurring during the Vertitalk Interval.
Source: Scientific Atlanta

An incorrectly set peak white clipper can be detected as compression of the peak video if set too low, or as permitting over-modulation if set too high. (Of course, this would best be checked during non-program times, as it would be necessary to intentionally attempt to over-modulate.)

Synchronous compression is another fault that could be checked using the test demodulator. If the synchronization and zero chops are set as above, and the blanking level is offset too low on the monitor, then synchronous compression may be present. Naturally, the fault may also be with the input to the demodulator, so this too must be checked.

Use of the demodulator also permits checking differential phase and gain in the modulator. Differential phase is a change in phase of the color subcarrier as the luminance level varies, and results in hue shifts as a function of brightness. Differential gain is a change in chroma amplitude as the luminance amplitude changes, resulting in errors in color saturation.

22.8 Noise Measurement

A demodulator is used in conjunction with a noise measurement test set to make in-service measurements of *signal-to-noise ratio* (S/N). The signal is

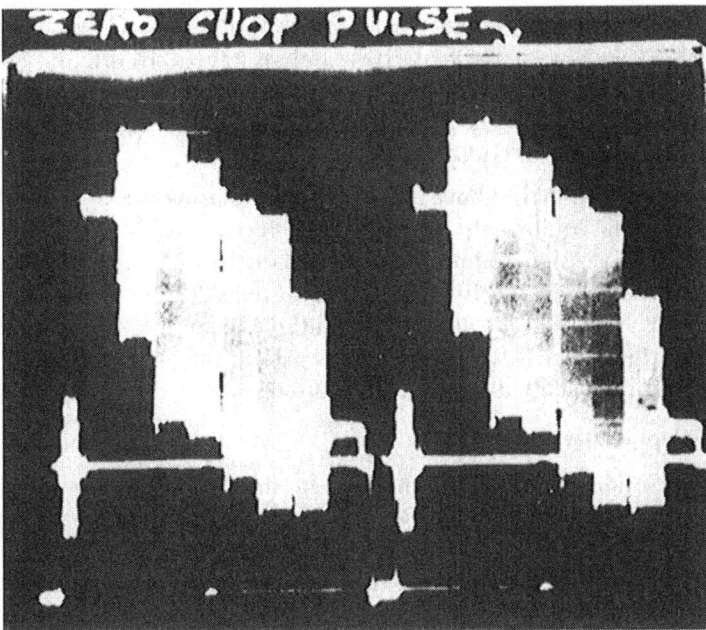

Figure 22.3 Demodulated Signal (Two-Line Display)
Source: Scientific Atlanta

demodulated and the baseband video is supplied to noise measuring equipment. The noise measured in this manner is the total noise accumulation between the point of clean video origination and the point of demodulation. Thus, the noise will include sources beyond the control of the CATV system. However, this total noise is the noise as seen by the subscriber on his home TV receiver. By making S/N measurements at the antenna, after the head-end, and in the distribution system, a realistic picture of noise contributions can be drawn, permitting the weakest points in the chain to be attacked.

In making S/N measurements, it is important to distinguish between different definitions of "signal" and "noise." In some definitions, the signal is considered to extend from the synchronized tips to peak video [EIA definition], while other, more common, definitions consider only the signal from blanking to white level [CCIR, BTL definitions]. Also, one of several weighting networks can be used, which roll off the higher frequency noise to account for its smaller contribution to subjective picture impairment. In all cases, "signal" is defined as the *peak-to-peak* value (with or without synchronization), m, and "noise" is the rms value.

These definitions may all be related to the National Cable Television Association (NCTA) definition of carrier-to-noise ratio (C/N). For example, the relation between NCTA-defined C/N and unweighted video S/N, where the signal is defined from blanking to white, is $S/N + C/N = 6.92$ dB, where S/N and C/N are both expressed in decibels and an RF bandwidth of 4 MHz is assumed. Were the signal to include the synchronization, the the S/N would increase by 20 log 1.7 = 2.93 dB, relating S/N to C/N by 3.99 dB. This is the 4 dB traditionally used in CATV work.

The relation stated in the above paragraph should prove useful in that video S/N measuring equipment, capable of in-service use, is available. This equipment defines signal as being from blanking to white, and may be used with or without a weighting filter. Note that the relationship between S/N and C/N is only valid if all the noise is attributable to sources that continue to exist after the carrier is removed obviously, and, this is not the case when noise is being transmitted from the TV station.

22.9 Antenna Alignment

With proper interpretation, a demodulator and wave form monitor can be used to aid in antenna alignment. In this application, only comparative readings may be used, because factors other than antenna alignment may affect results. These techniques are most effective if the station being monitored is transmitting *vertical interval test signals* (VITS). The photographs shown in this section were made with intentionally extreme signal distortions, generated by using a rabbit-ear antenna at different orientations.

Figures 22.4 through 22.7 show the VITS multiburst transmitted by a well run VHF television station in Atlanta. The receiving site was located about 10 miles from the transmitter. A high quality synchronous demodulator was used in making all of the pictures. In Fig. 22.4, the demodulator input was from a roof-top antenna oriented reasonably well. About 0.7 dB of peaking is observed in the middle of the multiburst. It is impossible to determine whether this is because of the transmitter, transmitting antenna, radio path, or receiving antenna. Of course, the demodulator could also be at fault, except that in this case its response was measured as flat with ±0.12 dB before these pictures were made.

The remaining photographs in Fig. 22.5 through 22.7 were taken with the same test set-up, except that rabbit-ears were substituted for the roof-top antenna. This was done intentionally to induce drastic changes. However, the same principles apply during antenna alignment.

In Fig. 22.5, the multiburst shows a roll-off of about 5 dB from the first burst. Re-orienting the antenna yielded about 7.5 dB of high frequency peaking, as shown in Fig. 22.6. Although some poor frequency responses are seen in these photographs, the pictures were all quite viewable to this point. In fig. 22.7, the antenna is again re-oriented, with very bad peaking as the result. This picture would probably not be considered viewable.

In Fig. 22.8 through 22.11 there is a similarly made frequency looking at the

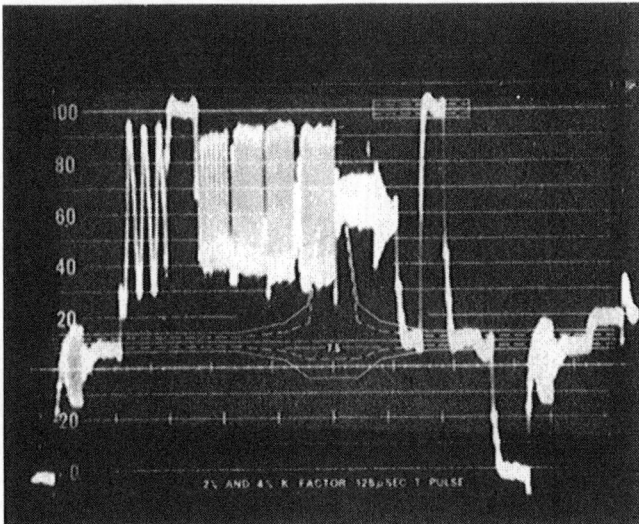

Figure 22.4 Demodulator Input from a Rooftop Antenna
Source: Scientific Atlanta

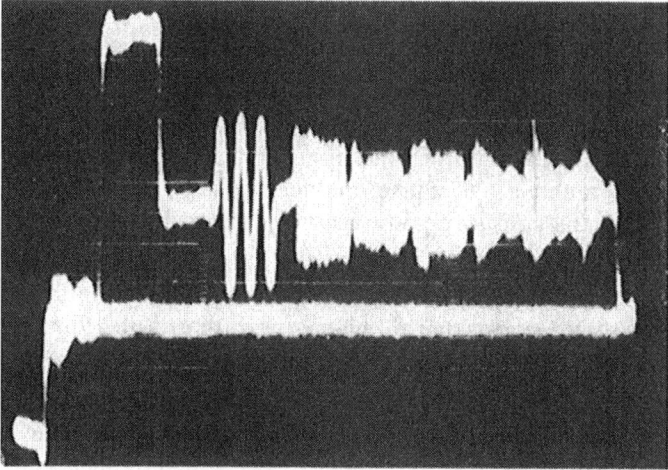

Figure 22.5 Demodulator Input from Rabbit-Ears (first burst)
Source: Scientific Atlanta

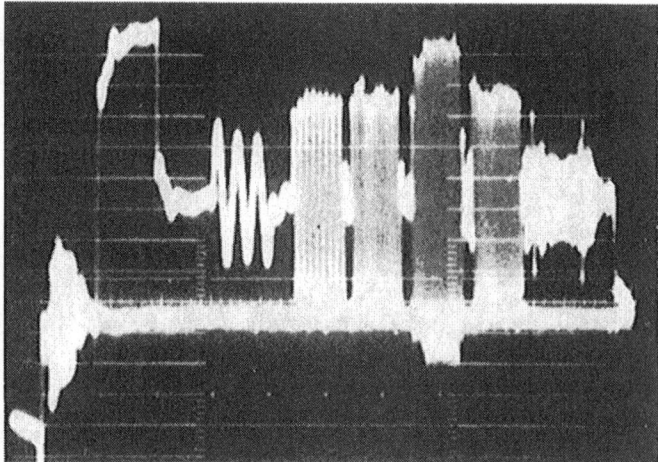

Figure 22.6 Demodulator Input from Rabbit-Ears (re-oriented, second burst)
Source: Scientific Atlanta

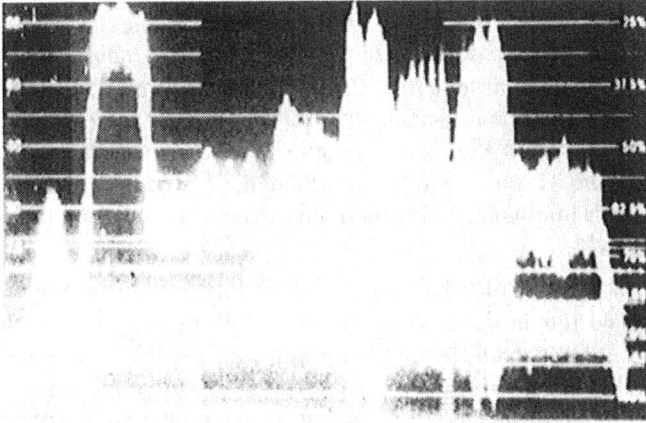

Figure 22.7 Demodulator Input from Rabbit Ears (re-oriented, third burst)
Source: Scientific Atlanta

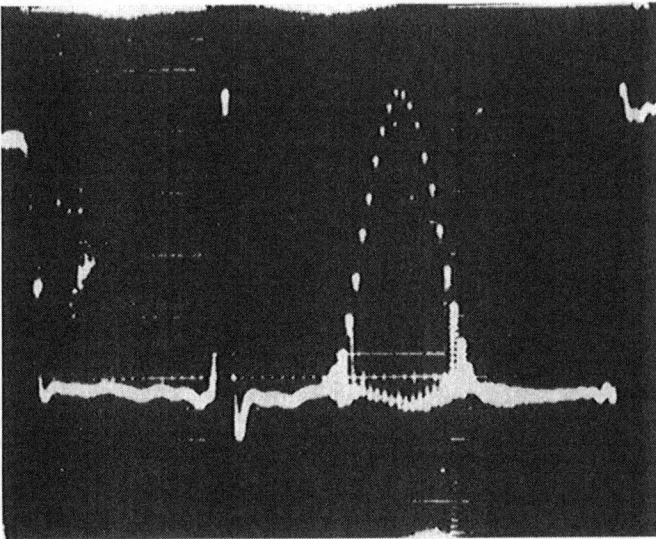

Figure 22.8 VITS (undershoot at trailing edge)
Source: Scientific Atlanta

2T and modulated 12.5T pulses transmitted in VITS. Figure 22.8 is from the roof-top antenna. This would be an excellent signal, with the only fault being the undershot at the trailing edge of the 2T pulse. Again, the demodulator was shown to have flat response prior to the test, but this undershooting could be caused by the transmitter, either the transmitting or receiving antenna, or the radio path. The strategy in using these tests is to orient the antenna for the best signal reception. In Fig. 22.9, two ghosts are seen in the two small pulses following the 2T pulse and the modulated 12.5T pulse shows evidence of considerable luminance-to-chrominance delay. Nonetheless, the picture was quite viewable.

In Fig. 22.10, the antenna is oriented such that the two ghosts are more pronounced and the delay is even greater. While degradation is quite perceptible in the picture, many people would still consider this an acceptable picture at normal viewing distances. Figure 22.11, on the other hand, was taken with the antenna oriented such that the picture is considered unacceptable.

22.10 Echo Quantification

The above discussion leads us to a method for quantifying echo (i.e., ghosting), one which should prove superior to conventional subjective efforts. For example, the first ghost seen in Fig. 22.12 and 22.13 is delayed from the main 2T pulse by about 700 nanoseconds (ns). A 700 ns delay would create a ghost displaced from the image by about 1% of the screen's horizontal width, or

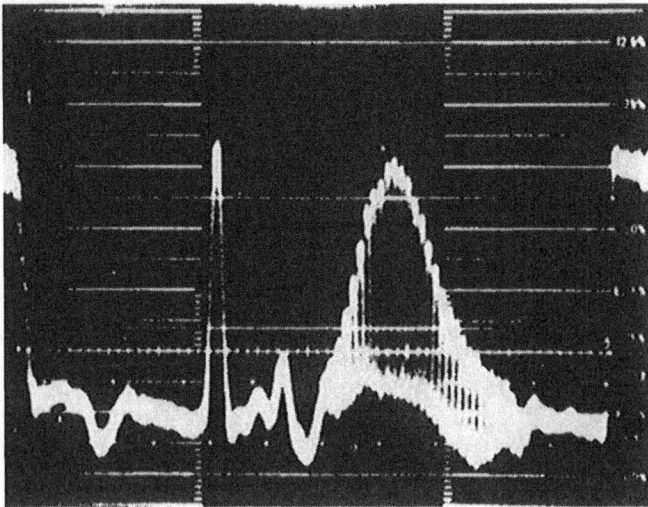

Figure 22.9 VITS (two ghosts visible)
Source: Scientific Atlanta

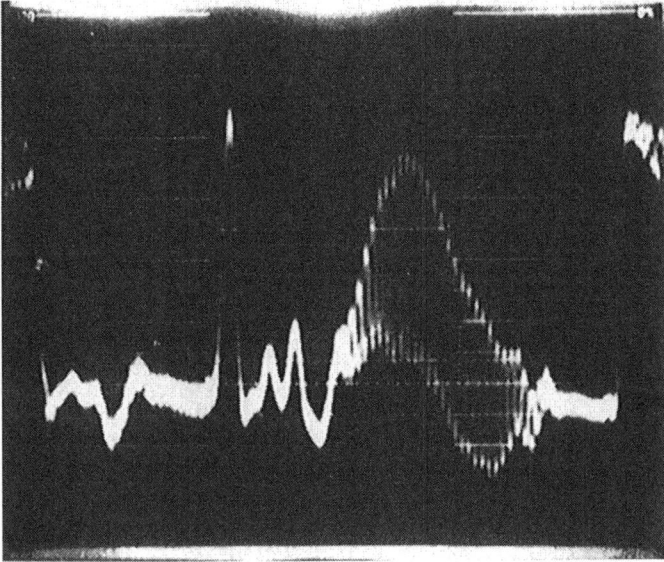

Figure 22.10 VITS (two ghosts more pronounced)
Source: Scientific Atlanta

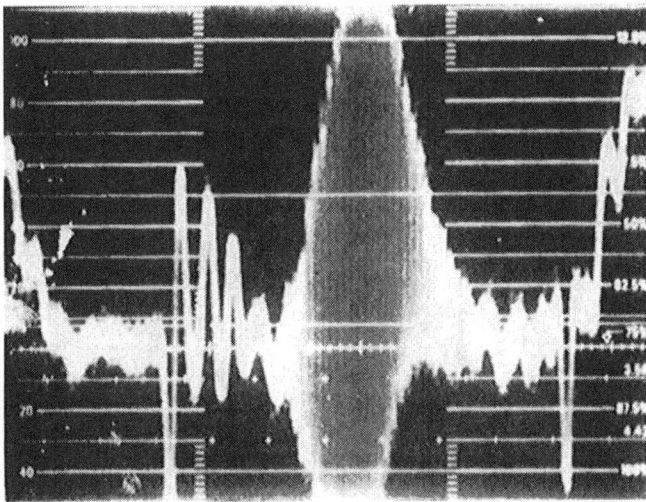

Figure 22.11 VITS (unacceptable picture quality)
Source: Scientific Atlanta

about 0.2″ on a 19″ screen. The relative amplitude of the ghost can also be measured. This first reflection is about 17 dB below the main signal. The Canadian Department of Communications provides a graph defining the visibility of an echo as a function of its delay and amplitude. According to this graph, an echo with 700 ns delay is considered visible if its strength is >32 dB below that of the direct signal. Thus, we would expect to see this ghost and the second ghost would be even more visible.

Because of the presence of the 12.5T pulse in the vertical interval, with about 2 microseconds (μs) between the two pulses. This technique would be limited to echo delays of less than 2 μs (about 0.6″ on a 19″ screen). For longer echoes, the bar signal which follows the 12.5T pulse can be used. This is shown in Fig. 22.12, where we have an echo delayed about 9 μs, and about –25 dB from the direct signal.

After concluding that echo exists, we must first rule out a problem with any other elements in the signal chain; i.e., transmitter, antennas, demodulator. For example, delay errors can cause shifts in the bar signal that might be confused with ghosting (although delay errors would probably not result in the extremely rapid transition to the final level, as seen in Fig. 22.12).

In all VITS use, the strategy ought to be utilization of the variables under

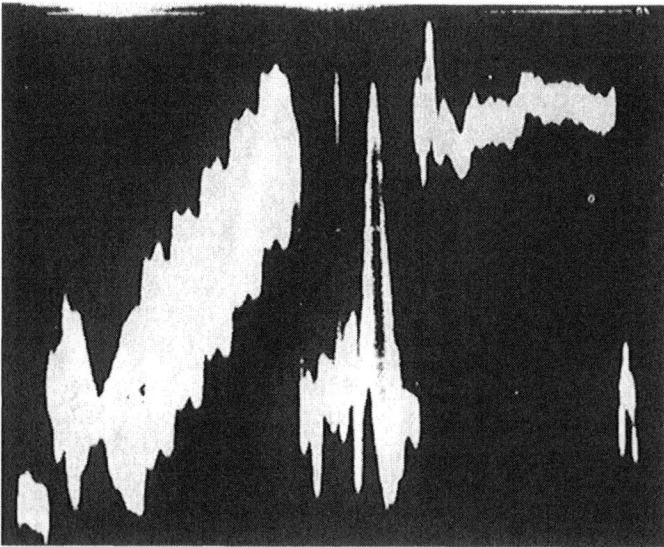

Figure 22.12 Echo Delay
Source: Scientific Atlanta

control (antenna height and orientation) to seek the best possible performance. The ultimate level of performance achieved, however, will vary considerably from one case to another.

22.11 Check the Head-End

Transmitted VITS may also be used to check head-end performance or an in-service basis. This can be done by comparing the VITS before and after the head-end. This does not substitute for normal proof of performance testing, but it can provide a useful estimate of processor performance without taking the processor out of service. Remember, the only comparison possible is between the VITS out of the processor and the VITS into the processor. The transmitted VITS will change, so incoming VITS must be checked before and after each use.

Head-end testing must be done with a quality demodulator in order to be truly meaningful. While the demodulated video from a TV set having this capability may provide an excellent picture, it is not satisfactory for measurement applications. The measuring techniques discussed are much more sensitive to signal degradation than is the picture. This is only logical because of the many sources of degradation encountered between the studio and the home TV receiver. If any one source of degradation were allowed to introduce noticeable determination in the picture, the overall degradation would quickly become intolerable.

22.12 Setting and Maintaining Modulation Levels

Two closely related problems are faced by users of television modulators: the problem of initially setting modulation and the problem of maintaining it over a long period of time. Indeed, there are four problems here, because the same problems exist for both audio and video. Failure to maintain correct video modulation may result in loss of synchronization, a washed out picture, and weak color in the case of under-modulation. Over-modulation can result in poor detector, amplifier, and kinescope linearity, causing problems with the intensity and tint of picture highlights in addition to excessive synchronous buzz in the audio channel. Under-modulation in the audio channel results in weak audio and poor signal-to-noise ratio. Over-modulation of the audio can result in distortion. Any errors in setting modulation level are particularly annoying when changing from one channel to another. Signal processors generally will not alter the modulation level of off-the-air stations, although frequency response errors will cause an apparent change in video modulation depth at higher video frequencies.

First, let us, consider the problem of setting (i.e., measuring) the video modulation level. Next we can consider the problem of maintaining it. Afterward, we will do the same for the audio level.

22.13 Measurement of Video Modulation

Video is transmitted by amplitude modulation (AM) onto an RF carrier. After modulation, one sideband is partially eliminated in order to conserve spectrum. The modulation format is such that synchronous tips correspond to the maximum carrier. Because of this arrangement, we refer to the depth of video modulation, defined as the percentage of the total amplitude change of the carrier, as the signal progress from synchronous tip to white. Standard video modulation requires that 87.5% = 12.5% of the maximum.

The simplest (and least accurate) method of determining video depth of modulation, is to display the modulator output on a TV screen, and adjust the modulation control until the picture looks good. Comparison with an off-the-air signal helps provide a reference. Unfortunately, this method is quite subjective and does not yield very accurate results. It should only be used in an emergency when no other method is available.

A second method of measuring depth of modulation is to connect a high frequency oscilloscope to the modulated IF signal and view the RF envelope directly. Under certain conditions, this yields a reasonably accurate measurement, but several circumstances can lead to incorrect conclusions. For example, if the spectrum of the IF signal being examined has been passed through a vestigial sideband filter to eliminate part of one sideband, then high frequency information (primarily color information) is not shown at the correct amplitude. This can be overcome by examining the signal before the vestigial sideband filter at the modulator, or after the Nyquist slope filter in the demodulator. Unfortunately, if harmonics of the IF signal are present at the point of measurement, these can render the oscilloscope display meaningless. Another drawback to the IF monitor approach is that the display presented is relatively difficult to interpret, especially in the case of a color signal. Additionally, this requires a rather expensive oscilloscope to obtain the required 50 MHz flat response.

The method of measuring video depth of modulation most often found in CATV practice is the use of a *modulation meter*. This is a meter that measures the peak-to-peak amplitude of a video signal, and expresses the result as depth of modulation. The video signal may be obtained prior to being applied to the modulation circuit, or it may be obtained by demodulating a sample of the modulator output. Although more expensive, the latter approach is preferred because accuracy will not be affected by variations in modulation sensitivity. Furthermore, by demodulating the signal for measurement purposes, a failure in the modulation circuit would be readily apparent.

A modulation meter is reasonably good for routine checks, but it has several drawbacks that limit its usefulness. For example, the frequency response of the metering circuit may not be flat over the entire range of the video spectrum. The meter depends upon the operation of a peak detector, which must accurately detect very short peaks and hold those peaks over at least one field of the picture. A fairly complex (and expensive) circuit is required to do this. Another drawback of using a meter to measure video modulation is that some types of distortion, which may accompany the modulation process, are masked. These include synchronous compression and incorrect operation of the peak white clipper.

The most satisfactory method of measuring video modulation depth is through use of a demodulator having a zero chopper. The zero chopper periodically switches off the picture carrier to simulate 100% modulation. This level and the synchronous tip level, representing 0% modulation, together provide the scaling required to measure depth of modulation. Figure 22.13 shows the output of a demodulator with a zero chop pulse generated during the vertical interval. A wave form monitor with a modulation depth graticule is used to display the output. By setting the synchronous tips on the 0% line and the chop pulse on the 100% line, the actual depth of modulation can be read directly from the scale. If such a graticule is not available, a regular oscilloscope may be used. If the synchronous and the zero chop pulses are set eight units apart, then each unit will represent a modulation depth of 15%. Normal modulation depth will then be represented by seven divisions.

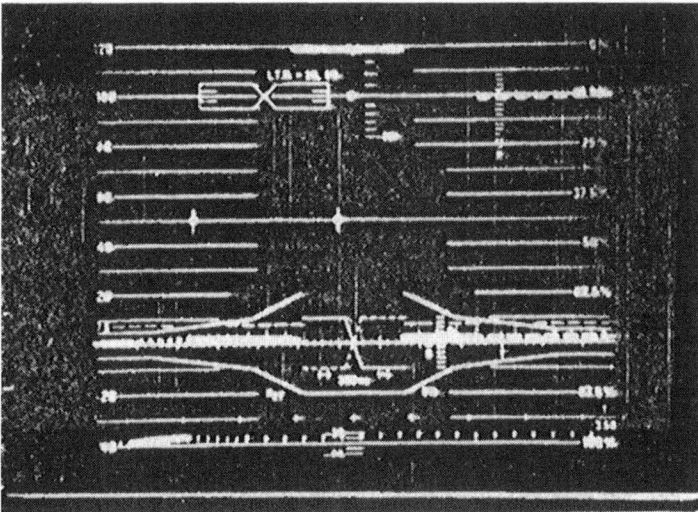

Figure 22.13 Demodulator Output with Zero Chop Pulse
Source: Scientific Atlanta

By measuring depth of modulation with a demodulator, other modulation faults, such as synchronous compression and improper operation of the peak white clipper, are apparent. Of course, the demodulator is also available to make many other measurements of modulator performance, as well as of off-the-air signals.

To avoid the necessity of purchasing input converters for each channel to be mentioned, the demodulator may be configured to accept an IF signal. The modulator IF output, taken just before the output converter, may be used for demodulator input.

Unfortunately, the zero chopper may itself exhibit errors in measuring depth of modulation. These errors are related to the non-ideal transfer function of the envelope detector diode. Designers have used several techniques to overcome this problem, including operating the diode at very high signal levels, switching a calibrating signal to the video amplifier during a chop, and biasing the detector diode. Detector problems may be overcome completely by using a synchronous detector, in which the injected carrier is used to overcome problems with the diode response. If an envelope detector demodulator is used for measuring modulation, then its response should be first calibrated using an alternate method of measurement.

Relative to the subject of measuring the depth of modulation, there is the question of whether the normal 87.5% modulation depth should include the color subcarrier. Often this is a moot point because most objects of high luminance have relatively small amounts of color saturation, resulting in inconsequential color subcarrier at the white level.

However, exceptions to this rule exist and, therefore, must be dealt with. One such exception, called the "Big Bird Syndrome," named after the bright yellow character of *Seasame Street* fame.* A case developed where trouble was observed in a broadcast transmitter, but only when Big Bird was on the screen. The problem was finally traced to a video processing amplifier which clipped the peaks of Big Bird's chroma signal, generating second harmonics that were not filtered by the vestigial sideband filter. These second harmonics caused adjacent channel interference.

Broadcast engineers disagree about whether to limit chroma peaks to 87.5% modulation depth. The author's opinion is that modulation should be limited to 87.5%, including chroma. This minimizes TV set differential gain and phase problems, and avoids having chroma amplitude reduced by the modulator peak white clipper with a corresponding drop in luminance level.

Sesame Street is a trademark of the Children's Television Workshop. *Big Bird*, Copyright © by Muppets, Inc. All rights reserved.

22.14 Maintenance of Video Modulation

Now that we have considered the measurement of modulation depth, let us turn to the problem of maintaining that modulation. If the amplitude of the incoming video signal remains fixed, and if the modulator is stable, then no problem exists. However, video levels may vary from time to time. The broadcast industry has developed several techniques for dealing with this problem, which find application under different circumstances. If the problem is one of maintaining constant output from one or more cameras in the studio, then it is desirable to monitor both black and white signal levels, forcing them to be consistent. This may be done with a video clamp to maintain a constant black level, and finding some means to vary the white level. White level adjustment might be accomplished at the camera by adjusting the iris, or it might be done electronically by varying video gain. Of course, the above signal must not include synchronization, and the synchronization must be added after adjustment. This technique is best left as a studio tool because it is inappropriate for some programs. For example, if automatic level correction were used on a night scene, the circuitry would attempt to make it a day scene.

If it can be assumed that at some point the ratio of video to synchronization has been properly established, and that this ratio has not been altered by subsequent video processing, then another technique can be used. The synchronous pulse amplitude is examined, and the amplitude of the composite video signal is adjusted to maintain proper synchronous amplitude. This will assure that the video modulation depth remains as intended, regardless of the maximum or minimum luminance level in a particular scene. This technique is sometimes applied at a modulator following a microwave link used to receive a distant signal. If the video amplitude provided at the microwave receiver output varies, consistent modulation depth could be restored.

A third method of control, used at some broadcast installations, involves tagging the problem with a reference signal at the point of origination. This *vertical interval reference signal* (VIRS) is theoretically transmitted through the entire transmission path, subjected to the same alterations as the video signal. It may be used to control several parameters of the transmitting equipment, including depth of modulation. In theory, this is an excellent (though relatively costly) technique, the author believes that it has not yet lived up to its promise for two reasons: the video amplitude is not always established correctly at the point of origin; and some point in the transmission path may inadvertently strip off incoming VIRS and retransmit a second VIRS, which is not necessarily related to the first.

22.15 Measurement of Audio Deviation

Let us now turn from the problems of measuring and controlling video modulation to similar problems in the audio channel. Again the problem is one of holding as close as possible to an established standard, in this case the FM deviation of 25 kHz, which should not be exceeded at any frequency.

If this deviation is exceeded, the signal will sound excessively loud and distorted. In addition, excess deviation can cause the sound carrier to rise out of the video path sound trap in the receiver. This, in turn, can cause the generation of 920 kHz beats in the picture, which will appear with every excess modulation peak. Under-deviation of the audio subcarrier will result in weak audio and a proper audio signal to noise ratio.

Several requirements are placed on the meter used to measure audio deviation. The first requirement is that the meter have good static accuracy; i.e., when a single tone is applied to the modulator input and the modulator is adjusted for an indication of 25 kHz deviation, at which point the carrier must actually be deviated 25 kHz. This can be established by measuring the sound subcarrier with a calibrated deviation meter. An alternative technique makes use of a spectrum analyzer. The modulator is supplied with a tone of known frequency, and the deviation control is adjusted until the spectrum analyzer indicates that the carrier amplitude has dropped to zero. The lowest deviation whereby this occurs is one for which the modulation index (ratio of peak i.e., deviation to modulating frequency) is 2.4. Because the modulating frequency is known, the deviation can be calculated.

As in the case of video measurement, the signal for deviation measurement may be taken either before the modulation process, or may be obtained by demodulating the audio subcarrier. The latter technique is preferred, as variations in modulation sensitivity will not affect accuracy. In either case, the signal used for measuring the audio should be given the same pre-emphasis bandwidth as the signal supplied to the modulated stage. Failure to do so will result in loss of accuracy if the highest amplitude signal component has a frequency greater than about 2 kHz. (With the 75 μs pre-emphasis time constant used in North America, the signal gain is raised by 3 dB at 2.122 kHz, and by about 17 dB at 15 kHz.)

Thus far we have concerned ourselves only with the *static* properties of the audio deviation measurement. However, the *dynamic* properties of the meter are quite important. Considerable effort has been expended toward the goal of measuring audio signal level in a manner such that all audio sources of equal measurement will sound equally loud to the listener. The conventional, widely used VU meter is one such attempt. However, it falls short of the goal. In measuring the deviation of the audio modulator, the goal is not only to meter sound for consistent loudness, but to achieve the maximum possible deviation without exceeding 25 kHz. This requires a meter whose dynamics are such that the peak level will be displayed.

Because this peak level exists for only a short time, an impractically fast response time is required of the meter. This means that the meter must be driven from an electronic circuit that compensates for the necessarily slow response of the meter movement. The required circuit rectifies the audio wave form and holds the peak value long enough to permit the meter movement to rise to indicate the correct value. This results in a meter movement whose characteristics differ considerably from those of a conventional VU meter. When modulation is applied, the meter begins moving toward the appropriate indication. When the modulation amplitude begins to drop, the meter follows, but sluggishly. A peak-reading meter has been adopted as the standard by the European Broadcast Union (EBU). In the United States, Schmid has compared its performance to that of a VU meter. He has found that the peak-reading meter permits much more accurate modulation monitoring.

One additional requirement placed on the dynamics of the audio modulation meter is that the overshot exhibited must be low. That is, when the meter reaches the final reading, it should stop quickly. All meter movements will exhibit some overshooting, but the amount should be minimized by proper meter selection and by properly matching the electronic driver.

The above dynamic properties may be explored on a given modulator by supplying the output of an audio signal generator (usually set to 400 Hz) to the modulator's audio input with a telegraph key in series. The key may be one of a number of hand keys used by amateur ratio operators for code transmission. When the audio is chopped into a series of short bursts (VITS), the meter should read the same as when the key is held down. When the key is suddenly depressed and held for a few seconds, the overshot exhibit by the meter can then be studied.

A somewhat more complex test was performed on several modulators sold to the CATV industry, and also on a distortion analyzer whose meter dynamics agreed with the standard VU characteristic. All modulator inputs were connected in parallel. An oscilloscope and the distortion analyzer were supplied audio through a pre-emphasis network. Tapes recorded from commercial radio broadcasts in Atlanta were used, and were spliced into endless loops so that the same segment could be studied on each indicator. First, a 400 Hz tone from an audio oscillator was supplied to all indicators, and levels were adjusted so that all modulators indicated 25 kHz deviation, the distortion analyzer indicated 0 dBm, and a reference trace was established on the oscilloscope.

The output of the tape recorder was then substituted for the oscillator. Only the recorder output level was adjusted in these tests, so all indicators received the same peak signal. For every test tape, the recorder output was adjusted for the same peak output, as indicated on the oscilloscope. On subsequent passes of the tape, each modulator meter and the distortion analyzer (with VU meter characteristics), were checked to see what deviations were indicated. The results are tabulated in Table 22.1, with errors in deviation translated to decibels.

Note first, from Table 22.1, the inconsistency in peak recording provided by the VU characteristic. This inconsistency would probably have been even greater had not all test tapes been recorded from processed audio. Often it is assumed that a VU characteristic meter will read about 8 dB below peak level on voice, with a somewhat lower error on music. Note next that modulators A

Table 22.1

Meter Response to Various Tests

Test	Modulator A	Modulator B	Modulator C	Modulator D	VU Meter
1	−2.8	−7.5	0	0	−7
2	−1.9	−5.7	+0.7	0	−6
3	−1.9	−5.7	0	−0.2	−4
4	−0.7	−7.0	0	0	−5
5	−1.5	−5.0	+1.3	0	−4
6	+1.9	−18.0	−6.4	−1.1	

The material on each tape is briefly described as follows:

Test 1—Male voice PSA, easy listening style.

Test 2—Heavy, explosive sound effects (from movie advertisement).

Test 3—Strong male voice.

Test 4—Male newscaster.

Test 5—Piano solo in higher octaves - fast notes with sustain pedal.

Test 6—Not really a tape, but included for comparison. This is a test of meter frequency response at 15 kHz. The signal level was reduced to take into account the pre-emphasis.

and B both read low on all tests. In neither case was the meter movement driven by a peak-detecting circuit. The meter movement in modulator A was sluggish, but exhibited overshot, which tended to make the readings appear to be a little closer to what should be expected. However, the overshooting can cause misleading indications in the opposite direction. In a preliminary test, this modulator indicated too high at times when the audio input was a female voice on a soap opera. The meter on modulator B exhibited less overshot, but it too was sluggish and failed to indicate true peaks.

The audio signal supplied to the meter circuit did not include pre-emphasis, as indicated by test 6. Either of these meters would be satisfactory if used only to set levels using a reference tone, with program level set elsewhere. However, only modulators C and D contain meters that would permit the correct deviation to be set using normal program material.

22.16 Maintenance of Audio Deviation

The final topic to be discussed in this chapter consists of the techniques used to maintain the maximum permissible deviation without exceeding 25 kHz. If the objective is to maintain the highest possible deviation, then automatic modulation adjusting equipment must take into account the pre-emphasis of the modulator. This is usually the reason for audio processing equipment designated for FM. The equipment first pre-emphasizes the signal, processes it, then de-emphasizes. Thus, the signal is processed with the same pre-emphasis that it will receive at the modulator.

Another precaution which must be observed when using audio processing equipment that is separate from the modulator, is the introduction of group delay between the processor and modulator. Group delay can shift the relative phase of different frequency components of a complex wave form, so that the peak amplitude of the signal could be changed after processing. This is of concern primarily to the stereophonic FM broadcaster who employs a 19 kHz trap in the input to the stereo exciter. However, this is only mentioned here as a precaution should a selective filter ever be necessary between an audio processor and a modulator.

Audio processing generally falls into one of three classifications: *compression*, *modulation limiting*, and *clipping*. Within each of the first two categories, there are several variations. These techniques and variations comprise the various audio processing approaches available. We will deal here with each, discussing suitable applications.

Compression is the term generally applied to the reduction of the entire dynamic range of the program material; i.e., "riding the gain" automatically. The objective is to maintain either a reduced dynamic range, or no dynamic range, in the output level. This technique is quite proper and popular for processing speech, where it can reduce such problems as level changes when a speaker turns away from a microphone, or when a new speaker begins talking. Compression is also used with rock music and other music formats in which dynamics are not a part of the art. However, extreme caution in applying compression is the rule with classical musicians whose work could have its necessary dynamics eliminated by over-zealous application of compression. Compression is generally specified by a ratio of N:1, meaning that N dB of input level change will result in 1 dB of output level change. A compression ratio of 2:1 would represent fairly small "meddling" with the dynamic range, while 10:1 would be a significant amount of compression.

Other criteria apropriate to compression include attack and release times, and a related question of whether the output peak amplitude should be monitored, or whether output rms or average level (taken over some time span) is to be maintained. A variation available involves equipment that separates the audio spectrum into two or more bands, compressing them individually.

Modulation limiting is the second technique for audio processing. Unlike a compressor, a modulation limiter does nothing to the audio level until a threshold (25 kHz peak in TV audio) is reached. Above this threshold, the limiter acts as a compressor with a very high (greater than 20:1) compression ratio. Thus, dynamic range is unaffected until a peak attempts to over-deviate the carrier. At this point, the gain is reduced until the modulation returns to a lower level. Modulation limiting is not to be confused with RF limiting, which clips the RF peaks, generating distortion. Modulation limiting is simply an automatic turning down of the gain if the level becomes too high. This technique is often used as a transmitter protection preventing over-modulation without introducing distortion. It is safe to use with all program material, if applied properly.

If the modulation limiter is over-driven, it acts as a compressor with very high compression ratio. This may be desirable in some instances, where a limiter is intentionally over-driven by a modest amount (8 dB seems to be a common figure) during normal programming. This permits minor decreases and increases in level to be compensated without masking the normal dynamic range of the material.

When using either a compressor or a limiter, levels cannot be set arbitrarily. There always exists a maximum level that can be accommodated by circuits prior to the gain adjustment stage. The difference in level between normal operation and the maximum signal that can be handled without distortion is called the "head-room." Head-room is specified for a constant sinusoidal tone, not for complex program material. If the rise in input exceeds the head-room for the particular equipment, then distortion will occur even if the output level does remain constant.

The final type of audio processing to be discussed is clipping, a process which cuts off the extreme levels of signal voltage. Clipping introduced distortion, and thus would not be used during normal programming. A clipper is normally set to act at some deviation greater than normal (e.g., 40 kHz for TV sound). It acts only to prevent gross over-deviation. If the signal were allowed to deviate too far, then the sound carrier would begin riding out of the sound trap preceding the video detector in the receiver, giving rise to 920 kHz beats in the picture.

Summarizing the three types of audio processing, a compressor is used for a range of audio problems, but if misapplied can be detrimental to the overall effort. Because the right amount of compression is a function of the type of programming, this is best left as a studio technique. The engineer concerned with signal modulation and transmission should concentrate on techniques which will not appreciably affect normal program audio. As a final clean-up measure, a clipper may be used to prevent accidental gross over-modulation, but this too should not affect normal program audio. When using a modulation limiter, a clipper is not as important as when no protection is supplied. However, because modulation limiters have finite attack times, a clipper could occasionally be needed to clip one or a few cycles of the audio, before the limiter acts.

In the final sections of this chapter, we have tried to mention some of the considerations apropriate to television modulator operation. While financial and operating considerations exclude some of the techniques discussed, the idea presented should be considered when problems arise or equipment upgrading is contemplated. Fortunately, no amount of electronic gadgetry and wizardry can replace a sharp technical person, who knows what techniques to use and where to apply them.

23 General Characteristics of Electromagnetic Waves

23.1 Velocity, Frequency, and Wavelength

A *wave* is any kind of oscillatory motion, the most familiar being waves on the surface of water. Sound waves, another example, are vibrations of the air or of various material substances. Both wave types involve mechanical motion. Electromagnetic waves are electrical and magnetic field variations which can occur in empty space as well as in material substances.

All waves are characterized by the property called *propagation*. The vibrations at a particular point in space excite similar vibrations at neighboring points, and thus the wave travels, or propagates, itself. The particular substance or space in which a wave exists is the *propagation medium*.

The speed at which an electromagnetic wave travels depends on the *dielectric constant* of the medium. The speed of electromagnetic propagation in a vacuum is of fundamental importance. This value is commonly called the *speed of light in a vacuum*, given as *c*. Light waves are actually electromagnetic waves of very high frequency. The value is 186,283 statute miles per second, or 299,793 kilometers per second, rounded off for most purposes to 186,000 miles per second, or $3 \cdot 10^8$ meters per second. The oscillations of waves are *periodic*. They are characterized by a frequency, the rate at which the periodic motion repeats itself, as observed at a particular point in the propagation medium. The frequency is expressed in cycles per second, a cycle being one full period of the wave. A single-frequency wave motion has the form of a sinusoid, or *sine curve*.

The *wavelength* of an electromagnetic wave is the spatial separation of two successive oscillations, which is equal to the distance that the wave travels during one sinusoidal cycle of oscillation. If the wave velocity is *v* meters per second and the frequency is *f* cycles per second, the wavelength in meters is

$$\frac{v}{f} \qquad\qquad\qquad\qquad\qquad\qquad\qquad\qquad (23\text{-}1)$$

As has been noted, v may have different values in different propagation media. When the value in free space is used in Eq. (23-1), the resulting value of λ is the free-space wavelength, sometimes denoted as free-space wavelength in inches, where f is frequency in MHz or megacycles (MC).

An electromagnetic wave has two components, an *electric field* and a *magnetic field*. Each component varies sinusoidally in time at a fixed point of space, with time period $T = 1/f$ seconds, where f is the frequency in cycles per second. Also, at a fixed increment of time, there is a sinusoidal variation in space along the direction of propagation.

Both the electric and magnetic components of the wave are "in phase" in space; that is, their maximum and minimum occur for the same z (see Fig. 23.1). They are also in phase in time at a fixed value of z. They are both directed at right angles to each other and the direction of propagation, a relationship that they always bear in free-space propagation.

23.2 Polarization

Linearly polarized means that the electric vector has a particular direction in space for all values of z; in this case, the x-axis direction. Thus, the wave is said to be polarized in the x-direction. In the actual space above the earth, if the electric vector is vertical or lies in a vertical plane, the wave is said to be *vertically polarized*. If the E-vector lies in a horizontal plane, the wave is said to be *horizontally polarized*. It is conventional to describe polarization in terms of the E-vector.

The initial polarization of a radio wave is determined by the antenna that launches the waves into space. The polarization desired is one of the factors entering into antenna design. In some applications, a particular polarization is preferable; in others it makes little or no difference.

23.3 Radiation and Reception

The term *antenna* is defined as a device for radiating or receiving radio waves. *Radiation* results when high frequency electrical currents flow under suitable conditions. The mathematical description of the relationship between electrical currents and their associated fields can be derived from Maxwell's equations, which state that *whenever electric current flows, a magnetic field is set up in the surrounding space. Any variation of this magnetic field will result in the creation of an electrical field.* Because the two fields always exist together — one cannot exist without the other, unless it is non-varying — this exchange between the two fields provides the mechanism by which waves are propagated in space at the speed of light. Whereas a current flowing in an antenna will

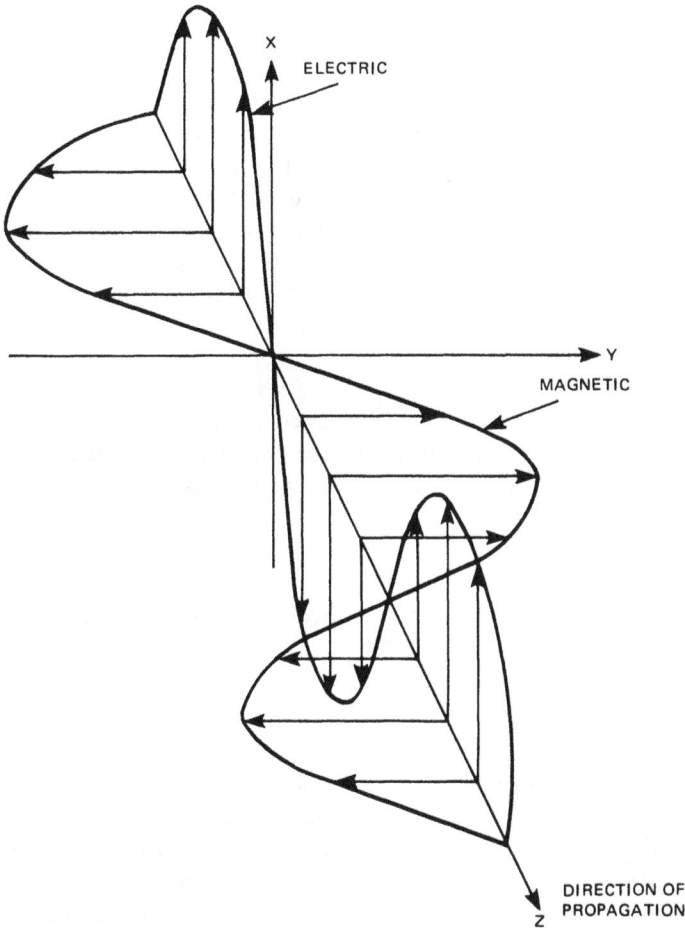

Figure 23.1 Two Components of Electromagnetic Waves
Source: National Cable Television Institute

produce radiation, an electromagnetic field cutting across on an antenna will cause current to flow in it. A *transmitting antenna* launches the electromagnetic waves into space. A *receiving antenna* captures energy from the incoming field and converts it into electrical current. Maxwell's equations also predict this effect, stating that *an electromagnetic field will cause current to flow in any conductor exposed to it. These induced currents will be of the same frequency, proportional amplitude, and phase variation as in the surrounding field.* It follows that these variations of current in a receiving antenna will be proportional to the variations in the transmitting antenna that launched the waves.

23.4 Reciprocity Theorem

The reception process is exactly the reverse of the transmitting process, any antenna can be used for either transmitting or receiving, the only restriction being that a transmitting antenna must carry heavy currents and be able to withstand high voltages. Antennas intended mainly for receiving may not have enough conductors or large enough insulators for transmitting. The currents and voltages in receiving antennas are ordinarilly so small that they are immeasurable with ordinary electric meters, i.e., they must be amplified. In every other respect, receiving and transmitting antennas are indistinguishable, and will work equally well for either purpose.

23.5 Log Pattern

Figure 23.2 shows the variation of the logarithm of power density, or electric field intensity, at a constant radius from the antenna as a function of angle. The logarithm of field intensity, E, in any direction can be expressed as:

$$20 \log_{10} = \frac{E}{E_{max}} \qquad (23\text{-}2)$$

The logarithm of the power, P, in any direction can be expressed as:

$$10 \log_{10} = \frac{P}{P_{max}} \qquad (23\text{-}3)$$

23.6 Antenna Gain

An *isotropic antenna* is one that radiates uniformly in all directions of space. Its pattern is a perfect spherical surface in space; i.e., if the electric intensity of the field radiated by an *isotrope* is measured at all points on an imaginary spherical surface with the isotrope at the center, or in free space, the same value will be measured everywhere. Such a radiator is not physically realized for coherent electromagnetic radiation. The concept of such an ideal omnidirectional radiator is, however, useful for theoretical purposes.

A *non-isotropic antenna* will radiate more power in some directions than in others; meaning it has a directional pattern. A directional antenna will radiate more power in its direction (or directions) of maximum radiation than an isotrope. (See Ch. 20, sec. 20.20.)

23.7 Directive Gain

The *directive gain* of an antenna is defined as a quantity that may be different in different directions. The relative *power density* pattern of an antenna becomes a directive gain pattern if the power density reference value is taken as the power density of an isotrope radiating the same total power, instead of using the power density of the antenna in its maximum-radiation direction as a reference.

23.8 Directive versus Power Gain

The previous definition of directive gain compares the power density of an actual antenna and an isotrope, based on the assumption that both are radiating the same total power. Another concept of gain, the *power gain*, compares the radiated power density of the actual antenna, and that of an isotrope, on the basis of the same input power to both. The isotrope is assumed to radiate all the power, but some of the power delivered to the actual antenna may be dissipated in ohmic resistance; i.e., converted to heat. The power gain takes into account the antenna efficiency as well as its directional properties.

The efficiency factor, k, is the ratio of the power radiated by the antenna to the total input power; it is a number between zero and unity. If the directive gain is denoted by D and the power gain by G, the relationship between them is

$G = kD$

If the antenna has no ohmic losses and, therefore, radiates all the power delivered to it, then $k = 1$ and $g = D$; the power gain and the directive gain are equal. At television frequencies, k is very nearly unity for a well designed antenna, so that the distinction becomes unimportant.

23.9 Gain Reference

The foregoing definition of directive gain is based on an isotrope as the reference radiator. Occasionally, the directive gain and directivity will be given relative to some other reference antenna, such as a *half-wave dipole*. In the past, this was the customary reference antenna. Present practice, however, is to use a hypothetical isotropic radiator as the reference antenna for computing directive gain and directivity. This may be assumed to be the reference unless otherwise stated. The following formula is useful in converting from either reference:

$G_{iso} = G_{dipole} + 2.15$ dB

Note that a dipole has 2.15 dB more gain than an isotrope; but, when converting, the given formula assumes that the isotrope is referenced to an isotrope and the dipole to a dipole. The formula is for reference changes, not comparing dipoles to isotropes.

23.10 Estimation of Directive Gain

Gain of an array $G = \epsilon D$, where ϵ is a number less than 1 that takes all losses into account other than those due to n. The aperture efficiency has already operated to broaden and so its effect is included in D. The number of square degrees subtended by all space is 41,253.

It should be pointed out that these calculations apply to arrays that produce only a single main beam. This includes a linear array of point sources whose main beam in the vertical plane has a beam angle of 360°, and it includes arrays that radiate into a half-space such as a planar array of dipoles over a ground plane. If the ground plane is removed so the planar array radiates a narrow beam in both directions, the D is divided by 2. In general, D is divided by the number of equal multi-lobes (main beams).

Theoretically, the directivity gain can be computed from the radiation patterns. This requires many patterns and graphical integration. Occasionally, it is of interest to obtain an approximate value for the gain when the only data available are the principle plane radiation patterns. Assume BW_E is the beamwidth between half-power points in the E-plane, and BW_H is the beamwidth between half-power points in the H-plane. Then

$$G = D\frac{41,253}{(BW_E)(BW_H)} \quad G = \epsilon D \tag{23-4}$$

where G is the power gain, BW_E and BW_H are in degrees. To express the gain in dB:

$$G_{dB} = 10 \log_{10} \frac{41,253}{(BW_E)(BW_H)} \quad G = \epsilon D \tag{23-5}$$

The value obtained in this fashion will generally be accurate to within ±2 dB.

23.11 Input Impedance

To connect the transmission line to the antenna, a small gap is made in the antenna conductor, and the two wires of the transmission lines are connected to the terminals of the gap — the antenna input terminals. At this point of connection, the antenna presents a load impedance to the transmission line. This impedance is the *input impedance* of the antenna. For a transmitting antenna, if the input impedance is equal to the characteristic impedance of the transmission line, there will be no *standing wave* on the line; otherwise, there will be. For a receiving antenna, if the input impedance is equal to the characteristic impedance of the transmission, all of the received signal will be transferred to the transmission line. Otherwise, some of the signal will be reflected back into the receiving antenna, and re-radiated. This re-radiated signal causes an apparent loss in antenna gain. It is interesting to note that the mismatch of the receiving antenna does not give rise to standing waves on the transmission line because the receiving antenna acts as the generator. Consequently, in order to generate standing waves, it is necessary that the load (i.e., preamplifier input or receiver input) be mismatched.

23.12 Bandwidth

Bandwidth, unlike the other parameters of an antenna, is one which does not have a unique definition. Depending upon the operational requirement of the system with which the antenna is to be used, the functional bandwidth of an antenna may be limited by any one or several of the following: change of pattern shape, increase in minor lobes, loss in gain, or deterioration of input impedance. In general, the only appropriate definition for the bandwidth of an antenna is *that frequency range within which the antenna meets a given set of specifications*. Many times, bandwidth is the angular distance between the half-power (3dB) points of a transmitted beam.

23.13 The Yagi Antenna

One of the most common antennas used in CATV is the Yagi antenna. This antenna consists of one or more driven elements, such as dipoles or folded dipoles, in conjunction with one or more reflectors and director elements. (See Ch. 20, sec. 20.8.) It is interesting to note that this antenna was erroneously named the Yagi antenna. This antenna was first described in Japanese by S. Uda, a professor at Tohoku Imperial University in Japan. Subsequently, it was described in English by one of his colleagues, H. Yagi. Because Yagi's paper was widely read among the English-speaking world, it has become customary to refer to this new development as the Yagi antenna. This occurred in spite of the fact that Mr. Yagi was quite clear about the part Professor Uda had played.

The Yagi antenna is a high gain compact antenna. The compactness is accomplished at the expense of bandwidth. Usually, the longer the Yagi antenna, the higher the gain and the narrower the beamwidths. The gain and beamwidths can be approximated by the following:

$$BW = \frac{55}{L\lambda} \qquad\qquad (23\text{-}6)$$

where

BW = beamwidth in degrees

L = the overall length from the reflector to last director measured in free space wavelength

and

$G = 13 \sqrt{L\lambda}$

or

$G_{dB} = 10 \log_{13} \sqrt{L\lambda}$

where

G = *gain as a ratio*

G_{dB} = *gain in dB*

A Yagi antenna must be designed to adequately cover the full TV channel. The bandwidth problem generally shows in the input impedance match and front-to-back ratio. Yagi antennas with multi-driven elements and compensating baluns have wider bandwidth than Yagis with only one driven element and that are gamma matched. Some Yagis that have 20 dB front-to-back ratios at one frequency in the TV channel may fall off to only 15 dB at another frequency in the channel.

The rate of performance degradation when the director elements are too long, is much more rapid than when they are too short. Hence, if the design is optimum at the picture carrier frequency, its performance will roll-off too fast to adequately cover the color and sound carrier. To cover the channel, it is advisable to make the director elements less than their optimum value at the picture carrier frequency. This roll-off in director performance also affects the front-to-back ratio. The front-to-back ratio may be improved by using a reflector screen behind the driven element; this will not be effective unless the screen is large enough. The minimum size is one-half wavelength square. Because of the mechanical support problem, it is difficult to realize a screen reflector for channels 2 to 6. However, screen reflectors work well on channels 7 to 13. The screen reflector and cantilever support of the element greatly reduce the adverse influence on antenna performance that the support tower or structure might otherwise entail.

23.14 Log-Periodic Elements

In the late 1950s, government research produced a breakthrough in state of the art antenna technology. This breakthrough was the concept of *frequency-independent antennas*, based on the theory that if a structure is made proportional to itself by scaling of its dimensions by some ratio, it will have the same properties at a frequency f and at a frequency τ. Therefore, the patterns and impedance of the antenna are periodic functions of the logarithm of frequency. By the proper choice of the properties of the periodic type of antenna the impedance will vary only slightly over the frequency band f to τ.

23.15 Corner Reflectors

The *corner reflector* antenna has been used successfully for receiving the high VHF and UHF channels. The corner reflector consists of a driven radiator, normally a half-wave element, associated with a reflector constructed of two flat conducting sheets, or their equivalent, which meet at an angle to form a corner: a 90° corner will provide 10 dB gain; a 60° corner nearly 12 dB; and a 45° corner about 14 dB. The power gain is produced almost entirely by compression in the H-plane. For this reason, the corner reflector does not give outstanding rejection to co-channel interference.

23.16 The Parabolic Reflector

An important geometric-optics concept, *focusing*, is familiar in connection with optical devices such as cameras, projectors, search-lights, telescopes, *et cetera*. Antennas utilizing radio wave reflectors focus rays in a similar way. If a beam of parallel rays is incident on a suitably curved reflector, the rays will be brought to focus at a point. Similarly, if a point source of radiation is placed at this focal point, the rays from it will be reflected in such a way that they emerge as a parallel beam; that is, the direction of all ray lines are referred. This is a simplified statement of the *principle of reciprocity*.

Rays that are parallel are said to be *collimated*; in other words, aligned in columns. Theoretically, they are focused at infinity. It can be shown that this collimation occurs if the shape of the reflector is *parabolic*. The automobile headlight is a familiar example of this type of parabolic reflector.

24 The Television Signal

The *television signal* is a composite result of many signals —horizontal scanning, vertical scanning, picture, color subcarrier, and aural subcarrier. This chapter will discuss these individual signals, how they interact, and present a graphical illustration of each of them. (See Ch. 22, sec. 22.1.)

24.1 Horizontal Scanning Interval

When we view a TV picture we observe the composite result of 525 individual lines of horizontal information. The *horizontal scanning interval* consists of the picture information for a given line, the blanking and synchronous pulses, and reference subcarrier (in the case of color video). One horizontal interval occurs in 63.5 μs, which is equal to a frequency of 15.750 kHz. Figure 24.1 details one line of horizontal scanning information. Blanking occupies approximately 10.5 μs of the time line. During this period, the electronic scanning beam in the TV receiver retraces from right to left without being seen in preparation to scan a new line. During the blanking intervals, a synchronous pulse of approximately 5 μs width synchronizes the TV receiver horizontal oscillator to the incoming signal horizontal frequency, and establishes the proper timing relationship between each horizontal line and the overall picture. Following the synchronous pulse on that portion called the "back porch" is a burst of eight to nine cycles of color subcarrier information (color video). In the generation of a TV picture the lines are scanned in two parts, called *fields*, each consisting of 262.5 lines. The two fields are then interlaced to create a single picture, called *frame*. One complete picture is generated in 1/30th of a second. More on this subject and line interlacing follows in sections 24.2 and 24.3, respectively.

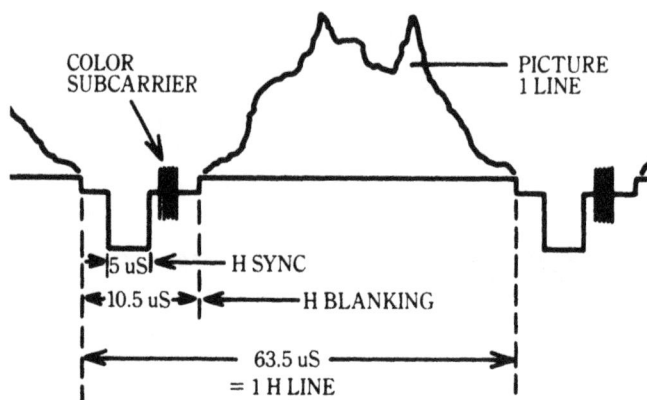

Figure 24.1 Horizontal Scanning
Source: Texscan Corp.

24.2 Vertical Scanning Interval

A complete TV picture is generated thirty times a second. The *vertical scanning interval* accomplishes the following during picture generation: synchronizes the TV receiver oscillator to the incoming signal's vertical synchronization, maintains horizontal oscillator synchronization during vertical blanking time, blanks out the picture scanning beam during retracing from the bottom of picture to the top, and causes *interlacing* of the two fields.

Figure 24.2 details one field of vertical scanning information. A field period of 16.6 ms is equal to the frequency of 60 Hz. The vertical blanking interval occupies approximately the first 1.16 ms of this period. To maintain horizontal oscillator synchronization during this period, a series of *equalizing pulses* related to the timing of the horizontal line rate are generated in the vertical interval. For approximately 190.5 μs prior to and following the vertical synchronous pulse, a series of six pulses at twice the horizontal rate are generated. Simultaneously, the vertical synchronous pulse is serrated into six parts with the same time relationship. This combination of pulses during the vertical scanning interval maintains horizontal oscillator synchronization. (More detail on equalizing pulses follows in section 24.3.)

The remainder of the vertical blanking interval is occupied with horizontal scanning lines, of which there are nine. A common practice is to transmit video test signals and data information on these lines.

24.3 Interlacing

One complete TV picture is produced by interlacing the two vertical fields. Figure 24.3 depicts the concept of field interlacing. Each of the vertical fields

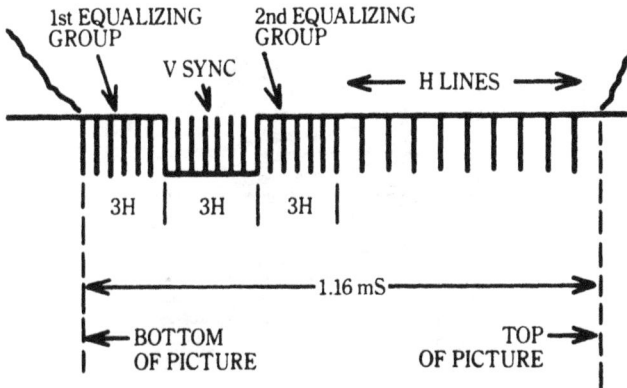

Figure 24.2 Vertical Scanning
Source: Texscan Corp.

contains 262.5 horizontal lines. Field 1, the odd field, begins scanning at the top left picture position and ends at the bottom right. Field 2, the even field, begins scanning at the top center picture position and ends at the bottom center. To achieve this condition, the odd field vertical synchronous pulse occurs one-half horizontal line earlier than the even field. To assure that proper interlacing of the vertical field occurs, the vertical synchronous pulse is preceded by six equalizing pulses which occur at twice the horizontal line rate, or 31.5 kHz. These pulses equalize any differential in the vertical synchronization integrator that would otherwise be present due to the field timing difference. As a result, the vertical oscillator instantaneously synchronizes, regardless of the presence of an off or even field. The six equalizing pulses following vertical synchronization maintain horizontal oscillator stability.

24.4 Color

The monochrome TV picture is created from a luminance signal which contains detail in the form of varying levels of grey scale; i.e., changes in black to white levels. [The foregoing discussion of color transmission is based on the standards of the National Television Systems Committee (NISC).]

To transmit a color picture we use the luminance signal to convey detail and add a subcarrier modulated by two signals containing *color hue* (phase) and *saturation* (amplitude) information. Also, short bursts of the subcarrier only are transmitted at the beginning of each line to phase-lock the TV receiver's color oscillator.

In order to transmit the color signal within the same bandwidth as the monochrome system, a technique called *frequency interleaving* is used.

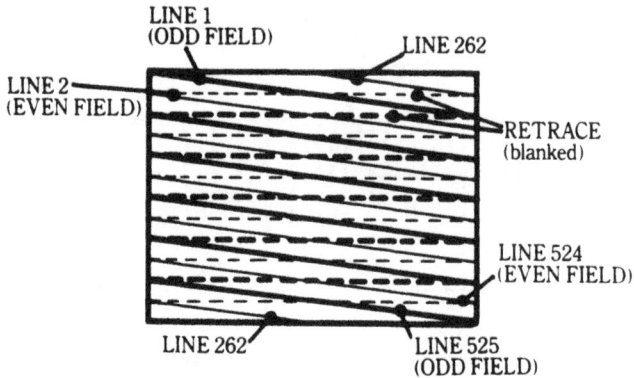

Figure 24.3 Field Interlacing
Source: Texscan Corp.

Due to the scanning method used in the TV system, sideband energy centers around even multiples of the horizontal scanning frequency. As mentioned at the beginning of this chapter, a complete picture frame contains 525 lines, an odd number. Due to the odd number relationship of line intervals from frame to frame, odd harmonic components cancel while even harmonic components add. In the gaps created by this phenomenon we interleave or insert, the modulated color subcarrier signal.

The frequency of the color subcarrier was chosen so as to interleave the luminance signal and the aural carrier (4.5 MHz above the visual carrier), while minimizing any beats between the carriers. The NTSC standards are: horizontal (line), 15734 Hz; vertical (field), 59.94 Hz; and color subcarrier, 3.579545 MHz (generally rounded to 3.58 MHz).

Picture details with frequencies above 1.5 MHz are produced by the luminance signal as brightness variations only. Details between 500 kHz and 1.5 MHz are produced by colors in the orange-cyan region and are transmitted as the *I* signal. Finally, details below 50 kHz are produced by three color combinations: red, green, and blue; and are transmitted as the *Q* signal. The *I* and *Q* signals modulate the color subcarrier. To eliminate any crosstalk between the *I* and *Q* signals, *quadrature* (90° apart) *modulation* is used. Figure 24.4 graphically illustrates the relationship between luminance, *I* and *Q* signals, and the resultant transmitted color subcarrier with *I* and *Q* modulation.

24.5 Picture Sound

The audio associated with a TV picture is transmitted by frequency modulation of a carrier which is 4.5 MHz higher in frequency than the visual carrier,

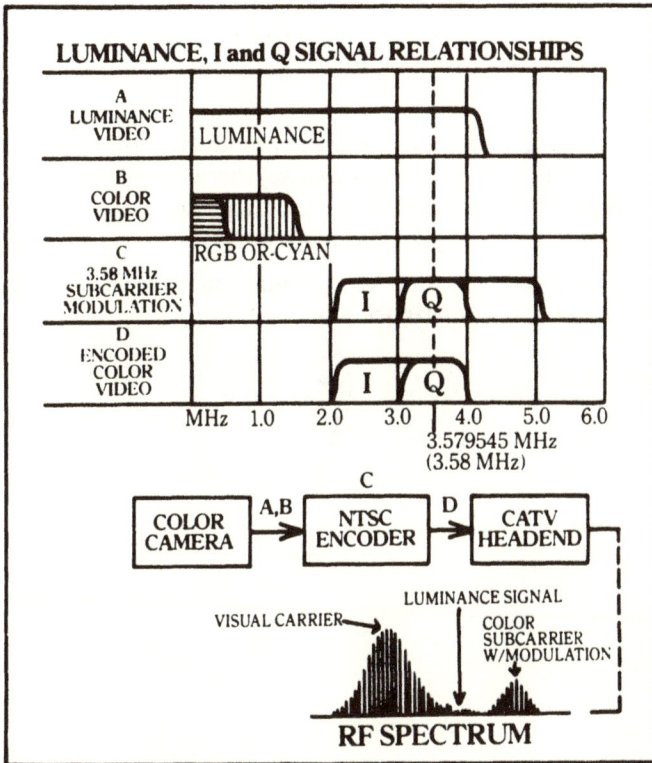

Figure 24.4 Luminance Signal Relationships
Source: Texscan Corp.

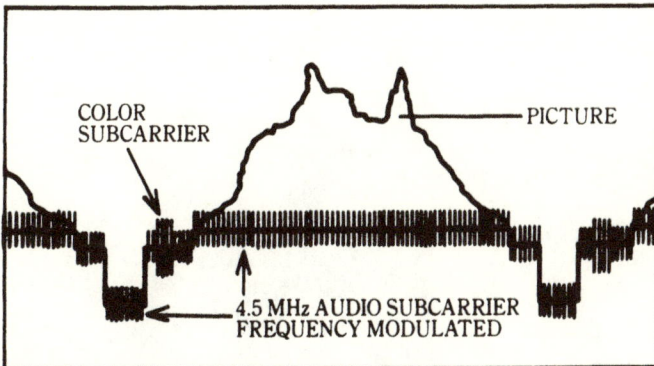

Figure 24.5 Video signal with Audio Subcarrier
Source: Texscan Corp.

or by frequency modulation of a 4.5 MHz subcarrier which is combined with the video signal. In either case, the TV receiver detects energy at 4.5 MHz from which the picture sound is recovered by *frequency modulation detection*.

Picture sound is interfaced to the frequency modulator through appropriate audio amplifiers and an *amplitude limiter*. The limiter is required to limit *peak deviation*, or the maximum carrier frequency variation, to the TV standard of 25 kHz, regardless of variations in the instantaneous audio peaks.

Figure 24.5 illustrates a horizontal line of composite video signal information with the 4.5 MHz audio subcarrier added.

25 Electronics Layout

25.1 System Parameters

The equipment chosen must be compatible with its physical environment. Where the atmosphere contains corrosive elements which pollute the air, such as long sea coasts or factories in proximity, jacketed cable may be necessary. All housings, connectors, and any other devices exposed to the elements should be protected from corrosion to a degree determined by the particular area. Equipment housings must also be mechanically sturdy enough to withstand the harshest weather conditions. Of particular importance is the watertightness of the equipment housing. In regions of high rainfall, water is probably the single most destructive factor to a CATV system.

In considering the compatibility of new equipment with older equipment in an existing system, if the numbers coincide such that the older amplifiers can remain, there are a number of equalizers and cable equivalents available which will allow the interface between the old and new system components to be equalized properly.

Another problem that frequently arises concerns the pilot carrier for AGC. If the pilot carriers are of different frequencies, the only requirement is to add the new pilot carrier at the head-end. If the pilot carrier is of the same frequency, and the same level is required, relative to the appropriate video carrier levels, there is no problem. However, if the pilot carrier is of the same frequency and a different level is required, a problem exists which can only be solved by trapping out the old system pilot and inserting a new pilot carrier of the proper level at the interface of the old and new system. Be aware that if equipment is properly selected beforehand, this problem can be avoided.

Once the equipment and cable have been selected, based upon published specifications, samples should be obtained from various manufacturers and these should be thoroughly tested to ensure that such equipment meets the published specifications.

Equipment evaluations should include electrical tests as well as environmental tests to check for corrosion resistance and watertightness. It is best to consult a local metallurgist to test for corrosion resistance. To guarantee watertightness, it is best to consult someone who has used that particular piece of equipment and who is familiar with its quality of operation under adverse conditions.

Once it has been determined that the equipment selected meets published specifications, these specifications can be used as the basis for system design. As a note of caution, remember that the equipment should be utilized only within the framework of its proven specifications.

25.2 Length of the System

The length of the physical plant plays an important part in the design of a CATV system. It is considered in terms of the number of cascaded amplifiers; this number determines the overall performance of the system. When the engineer is adding onto an existing system, including the spacing and types of amplifiers in that system, as well as the potential maximum length of the new addition, must be determined. The maximum length of the system is considered in terms of the minimum cascade possible to achieve that particular length of system.

The possibility of future expansion must also be taken into consideration. The length of such expansion should be estimated as closely as possible. Then, each particular cascade should be designed with a maximum cascade of the number of existing amplifiers, plus the number of amplifiers in the system or extension being constructed, plus an approximate number of amplifiers to be added at some future date. This allows future additions to be built without modifications to the existing system.

25.3 Temperature of the Area

The operation of a CATV system is greatly affected by air and radiation temperatures. The principal effect of temperature concerns the *attenuation* characteristics of coaxial cables. In reference to cable attenuation, three temperature figures are important: the annual average, the annual maximum, and the annual minimum, each based on 30-year records. The annual average is also often called yearly ambient temperature of the area. Cable manufacturers list their attenuation specifications at 70°F. If the annual average temperature is some other value, which is generally the situation, the cable attenuation must be corrected for the purpose of system design. For example, if a particular cable has an attenuation of 0.95 dB per 100′ at 70°F, at the ambient temperature of 50°F the cable's corrected attenuation would be as follows:

corrected attenuation

$$= \frac{0.95 \text{ dB}}{100'} + \left[\left(\frac{0.95 \text{ dB}}{100'} \right) \left(\frac{0.00127}{x° \text{ F}} \right) (50\text{-}70)° \text{ F} \right]$$

$$= \frac{0.95 \text{ dB}}{100'} - \frac{0.0217 \text{ dB}}{100'}$$

$$= \frac{0.9283 \text{ dB}}{100'}$$

$$\approx \frac{0.93 \text{ dB}}{100'} \tag{25-1}$$

The normal maximum and minimum temperatures must also be known because the total temperature variation determines the amount of AGC necessary to maintain the system within its specified tolerance. An extremely wide deviation might necessitate expansion loops in the cable. An extremely cold temperature will rule out some types of polyethylene jackets, as well as severely limit the available ampere hours from standby battery power supplies. Extreme heat will greatly affect the reliability of some transistorized devices. The worst case of signal-to-noise ratio will be a direct function of the minimum temperature.

Information concerning temperature variations is readily available at any weather bureau (NOAA) first-order station, and the state climatologist's office is a good source of assistance in interpreting data.

25.4 System Operating Parameters

Some suggested operating parameters are listed below. These parameters are by no means absolute. The numbers of any specifications can change, but the method for system design always remains the same.

Signal-to-noise ratio (for 4 MHz bandwidth and at maximum operating temperature)	40 dB
Cross-modulation ratio (at minimum operating temperature)	−52dB
Signal-to-hum ratio	40 dB
Minimum signal level at subscriber's set	3 dBmV
Maximum signal level at subscriber's set	15 dBmV

Although there are other specifications which must be taken into account, those listed above most directly affect the system layout.

25.5 System Maps

The *system map*, or strand map, is not only a geographic system diagram; it is the strategic layout for initial and future system development.

The location of all telephone and power company poles should be plotted accurately on system maps. Utility companies in a given area will have such maps and are usually willing to share them to the extent of allowing them to be copied.

The *field survey* is most important for preparing a system layout. After utility poles are plotted, the map is ready to have the field survey information entered. If the pole plot cannot be obtained from a local utility company, it can be done concurrently with the field survey.

The following items should be noted on the field survey:

1. Strand routing and easements used by the utility companies.
2. Poles where distribution is required, or likely to be required.
3. The maximum number of drops that may be required (100% subscriber saturation) on each pole where distribution is needed.
4. All multiple-dwelling units (duplex houses, apartment buildings, *et cetera*), and number of units in each.
5. All schools, television shops, or any other businesses which might require CATV.
6. All existing underground utility areas, including lot lines and obstacles to construction (trees, power transformers, *et cetera*).
7. Any pole positioned where it might be difficult to mount or service equipment, such as amplifiers, power supplies, *et cetera*.

25.6 Provisions for Future Service

After completion of the field survey, a visit to the local planning commission is in order to determine where and when any new development might be taking place within the system area. Several phone calls to local contractors might also reveal plans for new subdivisions or apartment buildings. The locations of these potential growth areas should be indicated on the system map along with brief notes supplying any available information, such as number of lots, dates of completion, *et cetera*.

25.7 Review

The object of this preliminary work is to provide a set of ground rules upon which system design can be based. Initially, the equipment is selected according to the area in which it must operate and the job it must do. Then, the equipment is thoroughly tested, both mechanically and electronically, to ensure that it will do what is required of it. In addition, system maps are drawn containing all of the information pertinent to the system design. This information is taken directly from the field and is vital to a good layout design.

25.8 Transportation and Distribution Trunk Design

Feeder lines are the connections between the distribution bridging amplifiers and the subscriber drops. Feeder lines may be bridged with line extender amplifiers and tapped. The feeder cables may be medium sized or large, but never as small as drop cables. Because the feeder lines often contain little or no active equipment, they are not always cascaded. Nevertheless, tap and line extender amplifier placement may cause some of the same problems and must be carefully considered.

25.9 General Procedures

The *trunk line route* is extremely important because it determines the total length of the trunk line. Before work on the layout is begun, it is smart to look at the system map and plot the head-end location and the shortest possible trunk line routes in order to provide signal to all of the required areas. This procedure coordinates the trunk routes with the overall picture of the system design.

The spacing between amplifiers, or between any two devices in a CATV system, should be thought of in terms of *decibels of attenuation*. There are not 2000 to 2100' between trunk line amplifiers, there are 20 to 21 decibels of attenuation between them. They may be 2000' apart, or they may be only 200' apart; it is the attenuation in decibels that determines the physical distance between amplifiers.

Relative to the physical distance between amplifiers, in the longest cascades it is advisable to stretch the amplifiers as far apart as possible for several reasons: first, to minimize thermal noise; second, to minimize distortion; third, to make the system more reliable by having fewer active components; and, fourth, to minimize maintenance. In keeping with this premise, it is desirable to tap the longest trunk lines, not split them, to form additional trunks. A two-way splitter has an insertion loss of approximately 3.5 dB. A high quality 16 dB hybrid directional coupler has an insertion loss, on the through side, of approximately 1 dB or less. If the directional coupler is used instead of the splitter to form a new trunk line off of a main trunk, the next main trunk amplifier can be moved forward another 2.5 dB, as compared to its placement using a splitter. The new trunk being formed is generally a shorter one, so the extra insertion loss will not hurt. Obviously, if the new trunk were to be as long as the main trunk, the splitter would be the best choice; but in most cases, this does not occur. It is not difficult to see that in a long cascade (20 to 26 amplifiers), it is possible to trim one or even two trunk line amplifiers off of the total cascade by tapping, rather than splitting, the main trunks.

All equalizers, pads, and cable equivalents should be included in the layout along with any special instructions about their installation that may be necessary. The layout should specify all components that make up the system and leave only fine adjustments to be done in the field. In effect, the layout should be a complete plan.

25.10 Cable Types and Applications

There is one cable available that is particularly well suited for underground application because it has a solid polyethylene dielectric, corrugated copper sheath, high molecular-density jacket, and a flooding compound for extra waterproofing. Foam dielectric cables, except those with solid aluminum sheaths, are not recommended for underground use.

Regardless of which cable is selected from the many different types available, the choice should first be based on the electrical characteristics of the cable, such as attenuation, return loss, shielding, and ability to handle the required amount of ac power; and, secondarily, on its mechanical characteristics which should be good enough to withstand the rigors of installation, water, extreme temperature, and any other environmental factors.

The two key factors of concern during system layout are *attenuation* and *slope*; assuming that all other electrical requirements are met. The cable slope is a direct function of cable attenuation, and must be compensated at each amplifier.

The cable attenuation (i.e. the cable size) is selected to provide the desired amount of length for a given amount of attenuation.

The attenuation and slope vary in a coaxial cable as a function of temperature. The system should be designed with the attenuation of the cable corrected for the ambient temperature of the area and attenuation changes due to temperature should be compensated by AGC.

25.11 The CATV Building Block: Unity Gain

In a CATV system, the input and output levels at all trunk line amplifiers should be uniform at ambient temperature. This is the most efficient mode of operation for a system of cascaded amplifiers. The losses preceding the amplifier plus the gain of the amplifier must equal zero. This means that there is no net gain and no net loss from the output of one amplifier to the output of the next. This concept of cascading like amplifiers is called the *unity gain concept* and ought to be used throughout the design of the system.

25.12 Determination of Operational Levels

The noise figure in a system of identical amplifiers operating at the same gain is

Noise figure in dBmV = NF_1 + 10 – log n

where

the system impedance = 75 ohms Ω
NF_1 = the noise figure for one amplifier
n = number of cascaded amplifiers

Figure 25.1 Noise Figure (identical amplifiers)

G_A = Amplifier gain
$-G_A$ = Cable attenuation
NF_A = Amplifier noise figure

Assuming a source noise across 75 ohms = –59 dBmV

Line D in Fig. 25.1 represents the output duration required as the number of amplifiers increases in order to maintain the same specified level of cross-modulation. The necessary duration is equal to 10 log n, where n is the number of cascaded amplifiers. This holds true only for a well behaved amplifier whose cross-modulation output decreases by 2 dB for every 1 dB that the output level is reduced. Most modern solid state CATV amplifiers are well behaved, according to the 2:1 output reduction ratio criterion.

25.13 Safety Margins for Variation in Temperature

The safety margin required in order to maintain the necessary signal-to-noise ratio and percentage of cross-modulation is a direct function of the maximum temperature deviations from the ambient temperature. The highest temperature encountered will determine the highest percentage of cross-modulation that is tolerable.

This can be reasoned empirically on the basic of the following principle: *as the temperature increases, the cable attenuation increases, which in turn decreases the signal level, while the noise generated in the amplifiers remains the same.* This results in a net decrease in the signal-to-noise ratio. Likewise, when the temperature decreases, the cable attenuation decreases and the signal levels rise. The output levels of all amplifiers without AGC will rise accordingly, thereby increasing the percentage of cross-modulation.

25.14 Trunk Line Layout

The trunk layout can now proceed on the basis of the unity gain concept. The system (strand) maps should have been completed, and all of the preliminary work should have been done. The equipment that is selected for the system will determine the operational gain, and it will be assumed to be 22 dB for the following examples.

The layout for trunk and feeder lines should be done at the same time. This is required to tie in the necessary bridger amplifiers and other distribution equipment. For our purposes, however, only the trunk layout itself will be considered in this section.

The first amplifier in the system is the natural starting point, so a few words are in order concerning the head-end and first amplifier interface. Most modern head-ends have an output level of between +50 dBmV and +60 dBmV per channel. Depending on the method of mixing, the length of this first spacing can be substantially longer than 22 dB.

In general, better overall system performance can be attained by using a highly efficient method of mixing and long first-amplifier spacing. Additionally, it is a good idea to leave a few decibels of reserve gain at the head-end to make the installation of future services, such as extra channels (or any one of a number of things that the future of CATV may hold), a simple addition to the head-end.

Continuing from the first amplifier, the layout is a matter of adding one building block (unity gain amplifier spacing) after the other. The main trunk lines should be routed throughout the city in such a manner as to minimize the maximum number of amplifiers in cascade. As mentioned previously, this is done by laying out the system maps and plotting a route from the head-end into each general area, checking carefully to be sure that the proper easements are being used. After the route has been sketched, the layout can proceed in an orderly manner from the head-end to the end of the line. Because the poles requiring distribution have been marked, the question of where bridger amplifiers will be required has already been answered. Additionally, it makes selecting the route of the trunk lines easier because any industrial area can be avoided, thus saving time and money.

Because poles are not spaced at exact intervals, some deviation from the recommended gain must be allowed. On a trunk line, this should be kept to within +0.5 dB and –1 dB. In other words, the amplifier should be overspaced no more than 0.5 dB, and underspaced no more than 1 dB. In practice, amplifiers can be underspaced by attenuating the input with a pad or cable equivalent. Once an amplifier is overspaced, however, signal degradation will result, which needless to say, it is not recommended.

25.15 Power Supplies

Contemporary solid state CATV systems are duplex-powered from centrally located power supplies. One such power supply may feed many amplifiers, depending on the transformer's rated load and the I^2R (power) loss in the conductors.

This method of supplying power to the active system components has many advantages. When vacuum tube equipment was being used, because of the larger power requirements, each amplifier location had to have an individual hook-up with the local power company. Electric bills were large; but, more importantly, reliability was tremendously reduced because of the reliance on the local power company in so many different locations. The use of transistorized amplifiers has greatly reduced power consumption and voltage requirements, so that one power supply can now feed a large portion of the system. In an efficiently designed system, the number of power hook-ups required can be cut to 5 or 6% of the number required in the same system using vacuum tubes. Alternating current is used almost exclusively because of the electrolysis problems encountered when using direct current.

Regulating power supplies should be used. They not only regulate the low voltage side, but they also provide excellent surge protection for the system. If the system is located in an area where lightning occurs, lightning arrestors should be used on each power supply.

25.16 Standby Power Supplies

Power supplies are available that have standby batteries and an inverter which can provide up to 36 ampere hours of reserve power in the event of a power failure. These standby power supplies have completely automatic switching and recharging. Therefore, they are almost maintenance-free. This type of power supply is highly recommended because it will save many hours of tracking down power outages and, more importantly, provide a much higher degree of reliability to the subscriber. Another bonus with this type of power supply is the additional surge protection provided by the delay in switching back to electric company power. A delay of at least 30 seconds, or preferably one minute, is desirable, because the secondary voltage can fluctuate widly for the first few seconds after the restoration of power service. A delay of one minute will ensure that these surges are over. It also gives the power company a chance to go off and on a few times, simply allowing the power company the chance to come to a steady-state voltage before the system is switched back onto it.

The use of standby power supplies requires consideration of the batteries themselves. Batteries are rated in terms of ampere hours of capacity at 77°F. For example, a battery rated at 36 ampere hours will theoretically run for 36 hours at 1 ampere or 1 hour at 36 amperes, or any combination of amperes and hours such that when multiplied together give 36 ampere hours. This is not exactly correct, because both the discharge rate and temperature of the battery affect the useable ampere hours. We shall use an example question and solution to illustrate.

Example: If a battery pack is rated at 36 ampere hours, how long can it provide full load current of 12 amperes at 0°F?

Solution: If the temperature were 77°F, then the discharge rate would be: 36 ampere hours/12 amperes = 3 hours.

25.17 Optimum Location of Power Supplies

Regulated CATV power supplies are available in a number of different ampere capacities. Most of these power supplies are of the *ferro-resonate transformer* type, which can regulate best when operated at, or near, full load. It has been shown that the voltage drop across the cable increases as the current increases.

When doing a system layout, it is most practical to do the complete electronics layout first, then do the power layout, taking advantage of trunk line splits to feed in several directions. Each branch of the power circuit is hence calculated separately with the sum giving the total power supply load. The general location of these power supplies should be decided, and then a field check made to determine their final location. This field check must be made in order to find a pole which has secondary power and available space to accomodate the power supply.

26 Underground Construction

With each passing year a greater percentage of new CATV systems are being installed underground. Because of this growing trend, it is essential that the cable television engineer be familiar with the basic installation techniques for burying CATV physical plant. The public relations problems peculiar to underground CATV installation are also of vital importance to the engineer.

The reasons for installing CATV physical plant underground include esthetic concern, system performance improvement, maintenance reduction, and economy. The majority of CATV operators who go underground do so in response to local laws. Often, people do not like to have additional poles in their alleys or along their streets — even for something as desirable as cable television. This attitude results in local laws requiring public utilities and CATV to be placed or replaced underground. In many newer communities the telephone and power utilities will already have been constructed or relocated below the earth's surface in order to remove poles, transmission wires, transformers, and guy wires from view. The result is a less obstructed view of the natural beauty of the community, but more work for the CATV operator.

In some communities, a CATV operator may elect to construct the physical plant underground because the power and telephone utilities, while still located on poles, have definite plans to relocate underground at some future date. Depending on time, it could be extremely uneconomical to attach CATV lines to utility poles which are already scheduled for removal. If a substantial amount of the aerial plant is scheduled to be replaced by underground plant within five years, for example, then the CATV system should definitely be installed underground.

Another reason for underground construction of CATV physical plant is the necessity for a good public image. If a substantial percentage of the towns-people strongly desire to have the CATV system strung underground, this concern should be carefully considered in deciding between underground or aerial construction.

A desire to improve system performance is often the determinant factor in electing to construct underground. With the cable and electronic components located in the ground, the effects of daily temperature fluctuations are almost entirely eliminated, center conductor pull-out is avoided, and, perhaps of greatest importance, degeneration and ultimate failure of components as a result of weather are avoided. Cable and amplifier life are often substantially prolonged because of their sheltered placement.

The threat of vandalism is also reduced, with underground construction, because an amplifier on a pedestal or in a vault is not much of a target for destructive mischief. Another type of vandalism, the "do-it-yourself" tap, is also virtually eliminated with keyed, locked, or bolted underground construction.

Another advantage of underground construction is ease of maintenance. Access to vaults and pedestals is relatively easy during almost any weather condition, especially as compared to climbing poles and using ladders and bucket trucks. Engineers and installers will not be required to have certain specialized skills, such as climbing poles and maneuvering a ladder or truck, which are essential to maintaining an aerial plant.

Although underground construction requires a higher initial cash outlay than an aerial plant, under most circumstances, there is some long term economic advantage to placing the CATV system underground. The cost of pole-line rearrangement or the construction of a private pole configuration is avoided, and the recurring cost of pole space rental is eliminated. In communities which will ultimately require all utilities and CATV lines to be placed under-ground, there is a definite monetary saving in initial underground placement.

26.1 Planning

As soon as the decision to go underground is officially made, a chemical analysis of the soil should be made. Soil conditions, particularly with respect to content, and the type and size of rocks should be considered. Soil samples from several locations should be collected and carefully identified as to location. The nearest Agricultural Soil Conservation Service (ASCS) office is probably the best source for soil testing. Independent test laboratories and colleges are also competent sources. Soil test results will indicate the type of soil and, thus, the extent of protection needed for the hardware which will be placed in the ground. To some extent, the soil test will also help to determine equipment selection. For example, the soil test will indicate whether metal pedestals can be used, because in some soil conditions a metal pedestal can be eroded extensively in less that a year.

Consideration should be given to the presence of underground rodents, and what precautions must be taken to protect the cable in areas of infestation. Information on gopher infestation is available from the U.S. Department of the Interior, Bureau of Sport Fisheries and Wildlife. Pocket gophers, the most prevalent cable chewers, are normally not a problem in heavily developed urban areas. Generally, if the cable is placed at a minimum depth of 24″, trouble from rodents can be avoided — even where a considerable gopher population exists. In addition, problems from frost heaving, soil bacteria, and accidental digging will be avoided by 24″ burial. Further protection is available in the form of specially compounded jacket materials which contain anti-rodent ingredients; i.e., ingredients which make the cable taste badly to rodents.

In virtually every case, a high-density polyethylene jacketed cable should be used for underground construction. This type of jacketing will suffice in all known soil conditions in the United States. If a conduit is to be installed — either containing the cable, or alongside the cable for future use — it should be made of PVC (polyvinyl chloride) if it is to withstand severe acid or alkaline soil conditions. For added protection, cables which employ a flooding compound have been proved effective. This material flows in a self-sealing process to fill accidental cuts and breaks in the outer polyethylene jacket which might occur during installation. Preliminary studies should also determine at what level the water table normally stands in various parts of the community. The presence of a very high water table will make the type of protection provided to components very critical. Many manufacturers fill electronic devices with polyurethene — a low-cost plastic material which completely waterproofs the device. Depending upon the type of enclosure used for electronic and passive devices, wide soil temperature variations may also create a problem which can dictate special protection for components.

In addition to evaluating soil type and condition, water table, and temperature variations in order to select appropriate equipment and protective materials, preliminary work should include an analysis of any special conditions which might affect construction or operation. These special conditions include the possibility of extremely heavy rainfall or extreme temperatures during the construction period.

26.2 Plant Layout

The accuracy of system maps is of the utmost importance in underground CATV construction. A facility which will be invisible when installed must be depicted accurately. Furthermore, the CATV facility will be located in precise relationship to other invisible facilities. Not only the maintenance of the CATV physical plant, but the safety of CATV personnel and the public depends upon the correct location of the coaxial cable facilities. This safety is contingent upon accurate mapping.

The overall system layout will be largely dictated by the available easements and rights of way. Generally, the CATV plant will be located in immediate proximity to telephone and power utilities if these have already been placed underground. In an area under development it is desirable to make arrangements with power, telephone, and gas companies for joint use of trenches. This will effect cost savings and help to ensure that proper clearances are maintained.

26.3 Trenching, Plowing, and Boring

There are two basic techniques for placing underground: the *open trench method* and the *plowing method*. Each has its advantages and disadvantages. *Boring* is only used for placing the cable beneath a structure or obstacle which cannot be cut or otherwise bypassed.

Cable plow installation is much more economical than trenching. The value of the cable plow is generally limited, except for planning house drops. The limitation of cable plow is primarily because of the existance of surface streets, surface alleys, curbs, gutters, *et cetera*. However, for long trunk runs, or for the installation of cable in undeveloped areas, a large cable plow pulled by a crawler can be extremely useful and economical.

The type of plow most commonly used in CATV construction employs the vibratory plow principle. For *direct placement* (cable which is placed directly in the ground without protective conduit), the cable is fed directly to the bottom of the cut through a cable guide from a reel which is mounted on the machine. With the *indirect placement* method, long continuous lengths of polyethylene conduit are pulled into the ground by means of a pulling grip attached to the plow blade. The cable is later pulled into the conduit. This method is recommended for installation of house (service subscriber) drops. Equipment which combines the advantages of small size and maneuverability is available for the job. Very little lawn disturbance occurs with this type of equipment, and low attenuation foam dielectric cable can be used without fear of damage during installation. Also, the conduit affords protection from accidental digging where the cable is installed beneath the lawn. The burial depth can be varied from 5″ to 12″.

There are a number of good trenching machines on the market. These machines have a very simple function — they dig a trench of specified depth and width, the bottom of which is normally level and relatively free from rocks and loose soil. The material removed from the trench is deposited in a uniform pile beside the trench, convenient for efficient back-filling.

If a construction contractor is hired for trenching, plowing, or boring work, it is important that the type of equipment as well as the exact specifications of the excavation be discussed with the contractor's supervising personnel before work is begun.

26.4 Open Trench Burial

There are two general methods of installation in an open trench: *direct cable burial* and *conduit* (or duct) *burial*. Direct burial installation is more economical than the conduit method. Direct burial results in savings in material costs of $800 per mile in addition to savings gained by eliminating the additional labor involved in conduit burial. However, the conduit method has several advantages over direct burial: (1) it provides a horizontal "hole in the ground" for future system expansion or cable replacement; (2) it protects the cable from damage during installation; and (3) it reduces the need for screening or for a sand cushion in rocky soil.

26.5 Pulling Through Conduit

Several factors must be given careful consideration when pulling cable into the conduit: (1) the cable must be swept before and after pulling to determine structural return loss and possible damage during the pulling process; (2) a dynamometer must be used to record the total torque placed on the cable during the pulling process; (3) seven hundred pounds of pull, as indicated on the dynamometer, must never be exceeded at any time during the pull. If the pull falls between 300 and 700 pounds, at least 3' of cable should be cut off on the pulling end and discarded before sweeping; (4) the center conductor must be firmly attached to a pulling grip before pulling any long lengths; (5) an approved wire pulling compound should be applied liberally on all long pulls; and (6) once a pull is started, it should not be interrupted until the entire length of cable has been pulled.

The keys to a successful pull are the proper placement of conduit, the firm attachment of the cable's center conductor to the grip, a liberal use of pulling compound, and a continuous, steady pull without interruption during the pull. If two or more cables are to be pulled into the same conduit, they should be pulled simultaneously rather than separately.

26.6 Installation of Electronic and Passive Devices

Installation of amplifiers, splice connectors, and passive devices must receive special consideration in buried physical plant. Many factors not normally associated with aerial construction arise when equipment is placed underground. Manufacturers are aware of these special requirements and have modified existing equipment or designed new hardware for burial.

26.7 Lightning Protection

In some areas of the US, the possibility of lightning damage to underground CATV plant is a matter of serious concern and should be taken into account during system design.

There are at least two different theories as to the precise nature of lightning-caused sheath puncturers in CATV cable. One is the *standing wave theory* which states that a surge is induced in the cable sheath when lighting strikes near an overhead section of the physical plant. The surge travels along the cable, and at points where the cable impedance to ground changes drastically, reflections are introduced. The combination of the traveling wave front and the reflections creates standing waves along the cable. At the peaks of these standing waves, the dielectric strength of the sheath is exceeded, the sheath is punctured, and the charge leaks off to ground.

A different theory suggests that the voltage is induced in underground cable by *ground current* resulting from a lighting strike. The changes in impedance to ground in this case are caused by variations in ground resistance or soil moisture content.

Regardless of which theory one accepts, there are three preventative measures. There is a transition point from overhead to underground, and special precautions should be taken at the last pole to obtain an adequate low-resistance ground. To stabilize the impedance of the underground section to ground, a bare copper wire (number 8 AWG soft copper wire is recommended) should be laid near the cable, preferably about 3″ directly above it. When the shield wire requires splicing, it should be spliced with a brass sleeve using a sleeve presser. The shield wire must be connected to the outer sheath of the cable at all amplifier locations and at other housings which are grounded.

26.8 Public Relations

Where the CATV physical plant is being installed in backyard easements, fences are a primary public relations concern. The ideal approach is to hire a professional fence company to drop the fences in advance of the crew and immediately replace the fences after the trencher has gone through. One cable operator found that it was more economical to temporarily drop five fences in any block than to take the trencher in and out of each individual backyard. This latter process of course, involves considerable maneuvering of the trencher to position it close to the fence and in alignment with the trench. It should be noted that sections of many fences will have to be totally replaced following this procedure. This method can still be more economical than the time-consuming process of going in and out of each yard with the trencher, and then tunneling under each fence.

27 Satellite Communications

The purpose of this chapter is to present an introduction to satellite communications insofar as a basic understanding is pertinent to CATV. There are many applications and aspects of the theory of satellite communications which will not be covered here. The approach taken here is to describe a simple FM-video link, as one example, using it as a vehicle to describe the background for the various link equations. The Glossary at the end of this book defines terms which occur frequently in the field of satellite communications.*

27.1 Description of a Simplified Satellite Link

The satellite is assumed to be at a longitude of 90° W in a geosynchronous orbit at a height of 35,800 km above the equator (see Fig. 27.1). The transmitting earth station is near New York City at 74° W, 41° N; the receiving earth station is near Los Angeles at 118° W, 34° N. It can be shown that the *satellite elevation angle* is 39.84° above the horizon, as viewed from the transmitting earth station, and lies generally to the south-southwest at an azimuth of 203.61°. From the receiving earth station, the satellite elevation angle is 40.44° and the azimuth is 136.44°.

Figure 27.2 is a block diagram of the communications link. It consists of a transmitting earth station, the satellite, a receiving earth station, and the propagation paths traveled by the signals. For simplicity, the satellite is assumed to have only one *transponder*, and the *uplink* and *downlink* frequencies are assumed to be 6 GHz and 4 GHz, respectively. In practice, there are no

---

*See D. Jansky and M. Jeruchim, *Communication Satellites in the Geostationary Orbit* (Dedham, MA: Artech House, 1983).

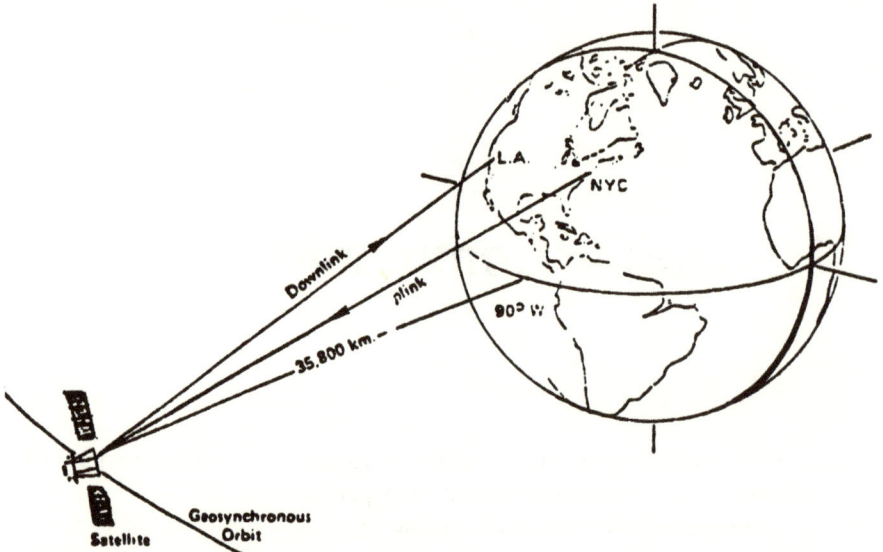

Figure 27.1 Satellite Link Geometry
Source: Scientific Atlanta

carriers operating at exactly 6 GHz, and in the 6/4 GHz band* downlink carriers are offset 2.225 GHz instead of 2 GHz below the corresponding uplink carriers. This simplified description assumes that only one earth station can access the transponder at a given time.

The uplink consists of the earth station transmitter and its antenna, the uplink propagation path, the satellite receiving antenna, and the transponder receiver. The satellite transponder receives the incident signal, converts it from the 6 GHz band to the 4 GHz band and transmits it back to earth. The downlink consists of the transponder transmitter, the satellite transmitting antenna, the downlink propagation path, and the earth station receiver antenna and its associated receiving equipment.

The quality of any communications link is determined by the difference between the output signal and the input signal. In an analog system, such a FM-video, the difference is represented by distortion and noise. In a digital system, it is represented by the *bit error rate* (BER). The satellite link under consideration must adhere to distortion and video signal-to-noise criteria which are appropriate to the application. In this simplified system description we hypothesize a 54 dB, clear weather, video signal-to-noise ratio at the receiving end.

---

*The designation 6/4 GHz is standard terminology used to indicate that the uplink is in the 5.925 to 6.425 GHz band and the downlink is in the 3.7 to 4.2 GHz band.

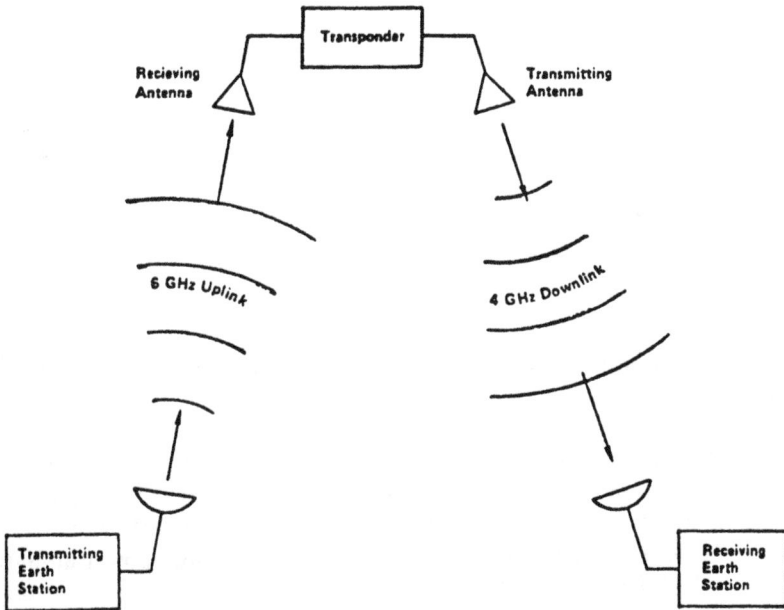

Figure 27.2 Simplified Block Diagram of a 6/4 GHz Satellite Communications Link. (Only one of the 12 or 24 transponders that make up a typical satellite is shown.)

Source: Scientific Atlanta

27.2 Uplink

Figure 27.3 is a simplified block diagram of the transmitting earth station. It consists of a frequency modulator, an up-converter, a high power 6 GHz transmitter, and a transmitting antenna.

Before modulation, the video signal is processed to pre-emphasize the higher frequency components, and an energy dispersal waveform is added. Pre-emphasis acts to improve the output video signal-to-noise ratio (S/N), by compensating for the increase in noise density with frequency (*triangular noise*), which is characteristic of the receiver discriminator. (The pre-emphasis is removed by a de-emphasis network after the receiver discriminator.) The energy-dispersal signal frequency modulates the carrier with a triangular waveform at the video frame rate to disperse the RF spectrum. This reduces interference with terrestrial microwave and other satellite links, and reduces intermodulation among the multiple carriers, which exist in a real satellite.

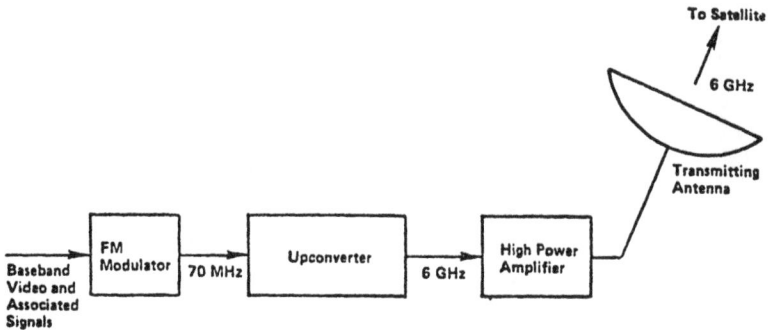

Figure 27.3 Simplified Block Diagram of Transmitting Earth Station
Source: Scientific Atlanta

The baseband video, and the associated audio and energy-dispersal wave-forms, are frequency modulated onto a 70 MHz carrier. The modulated signal, whose modulation bandwidth is approximately 36 MHz, is converted up to 6 GHz, amplified, and used to drive a klystron high-power amplifier (HPA), which feeds the transmitting antenna.

The transmitter and transmitter antenna are characterized by an *effective isotropic radiated power* (*EIRP*), given by

$$EIRP = P_o G_T, \text{(watts)} \tag{27-1}$$

where P_o is the transmitter output power in watts and G_T is the transmitting antenna power gain. Equation (27-1) can also be expressed in decibels by

$$EIRP = P_o + G_T, \text{(dBW)} \tag{27-2}$$

where *EIRP* and $P_o + G_T$ is in dB.

The transmitting antenna must be relatively large in diameter to generate a narrow beam. The narrow beam pattern protects other satellites from inter-ference from the transmitted wave and at the same time produces high gain, which is required to overcome propagation losses. The gain g (θ, ϕ) of an antenna defines the beam pattern, where θ and ϕ are angular spherical coordinates. The term *gain* used alone implies the gain at the peak of the pattern. In general, the higher the gain, the narrower the pattern.

The link in this simplified description uses the 10 meter diameter antenna of Fig. 27.4, which has a gain of 53.5 dB and 6 GHz. We will assume an *EIRP* of 80 dBW (10^8 watts), which is obtained from the 53.5 dB antenna gain and 26.5 dBW (about 450 watts) transmitter power.

On passage through the atmosphere along the uplink propagation path the *EIRP* is decreased slightly by atmospheric attenuation, which is very small at

Figure 27.4 Scientific-Atlanta Earth Station Using 10 Meter Diameter Antenna
Source: Scientific Atlanta

6 GHz, even in heavy rains (see Fig. 27.5). On the other hand, a very large attenuation occurs because of the spreading of the spherical wave front as the wave radiates outward.

The attenuation because of spreading can be seen from the definition of EIRP, which is the power that must be transmitted if the radiated power is to spread out uniformly in all directions from the source. The power density S at the satellite is given by

$$S = \frac{EIRP}{4\pi R^2} k_A, \text{ (watts / meter}^2) \tag{27-3}$$

where R is the distance from the earth station to the satellite in meters, $4\pi R^2$

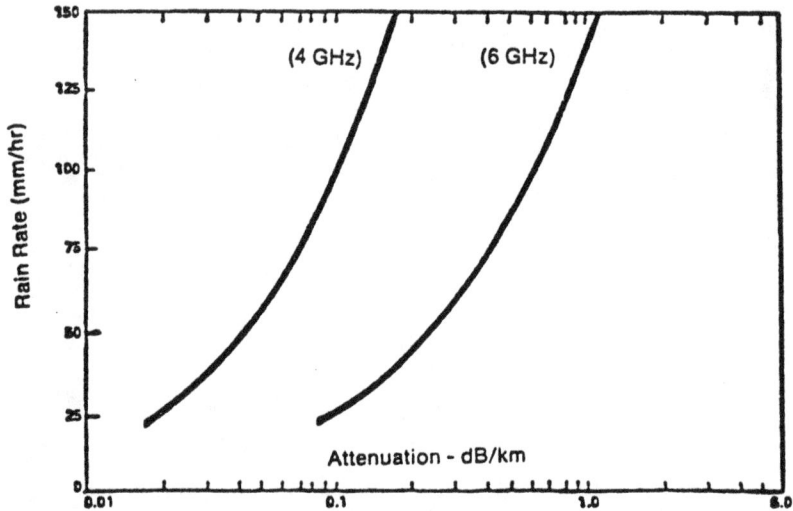

Figure 27.5 Attenuation vs. Rain Rate
Source: Scientific Atlanta

is the total surface area of a sphere of radius R, and k_A is the atmospheric attenuation factor.

The total atmospheric attenuation is obtained from the integrated effect of the attenuation over the total uplink propagation path. The total path which contributes to rain attenuation is relatively short because of the slant path to the satellite. Typical values of attenuation because of rain, even heavy rains, are likely to be less than 1 dB at 6 GHz (1 dB of attenuation corresponds to k_A = 0.794 in Eq. (27-3).

Contemporary 6/4 GHz satellites have either 12 or 24 transponders, each of which have bandwidths of about 36 MHz. Figure 27.6 is a simplified block diagram of a single-transponder satellite.

To accommodate signals received from widely dispersed geographical areas, the satellite receiving antenna has a broad-beam pattern. Figure 27.7 is a "footprint" showing contours of flux density which are required to saturate the satellite transponder.

At the start of our simplified description we assumed an *EIRP* of 80 dBW. We will now show how we arrived at this *EIRP*. Solution of Eq. (27-3) for *EIRP* and conversion to decibels give

$$EIRP \, (dBW) = 10 \log (4 \pi R^2) + S(dBW / m^2) + L_A \, (dB) \qquad (27\text{-}4)$$

Where L_A (= $-10 \log k_A$) is the atmospheric attenuation in decibels.

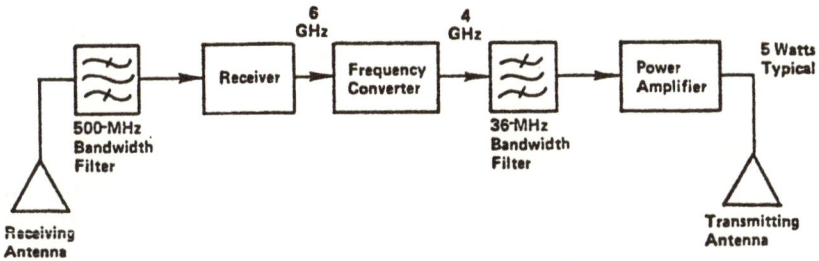

Figure 27.6 Block Diagram of Simplified Single-Transponder Satellite
Source: Scientific Atlanta

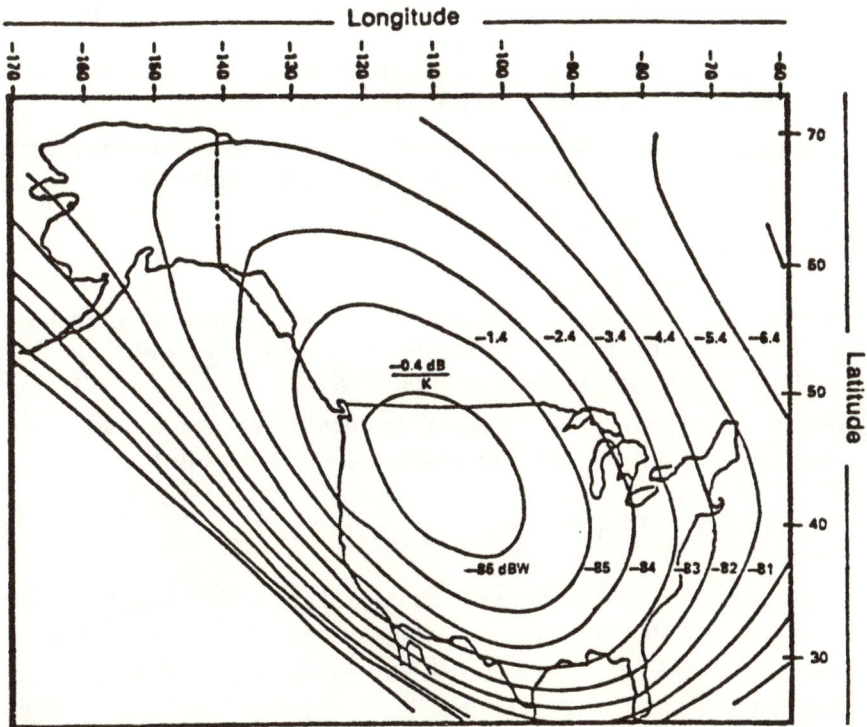

Figure 27.7 Typical Variation of Saturation Flux Density and G/T for a
Domestic US Satellite
Source: Scientific Atlanta

It can be shown that the distance from New York City to the satellite is approximately 37,750 km. If we use a saturation flux density from Fig. 27.7 of -82.5 dBW/m^2 for an earth station near New York City, and assume an atmospheric attenuation of zero dB, we have from Eq. (27-4):

$$EIRP = 162.5 - 82.5 = 80.0, (dBW) \tag{27-5}$$

This confirms that our assumed value of 80 dBW is nominally correct. In practice, several decibels of additional power would be made available by use in a larger HPA to overcome waveguide and switching losses, antenna pointing error, atmospheric attenuation, and other losses.

The power received by the satellite antenna is obtained by multiplying the incident power density in watts/meter^2 by the effective area of the receiving antenna in square meters:

$$C = SA_e\, P = \frac{EIRP}{4\pi R^2}\, k_A\, A_e\, P, \text{(watts)} \tag{27-6}$$

where the C designates carrier and p is the polarization efficiency (virtually equal to unity in a properly designed and adjusted system).

The effective area A_e implies a beam pattern, as does gain. The beam pattern for a given effective area is determined from the relationship

$$A_e\,(\theta,\, \phi) = \frac{G(\theta,\, \phi)\, \lambda^2}{4\pi} \tag{27-7}$$

where λ is the wavelength and where θ and ϕ are spherical coordinates describing directions from the satellite. These directions translate into the latitude and longitude of the footprint of Fig. 27.7.

Knowing the effective area of the satellite antenna, one can calculate C from Eq. (27-6), but it is customary to express Eq. (27-5) in the form

$$C = EIRP\, k_A\, G_R \left(\frac{\lambda}{4\pi R}\right)^2 p \tag{27-8}$$

by using Eq. (27-7) (G_R is the maximum value of $g\,(\theta,\, \phi)$). This is done because antennas are customarily specified in terms of gain rather than effective area.

Before considering the received carrier power C further, we will discuss noise power and carrier-to-noise ratio.

At the satellite, a certain amount of electrical noise power N_A enters the satellite receiver via the antenna along with the carrier power C. An additional amount of noise power N_R is generated in the low level stages of the receiver.

The noise power is uniformly distributed in frequency, and the noise power of concern is that which is contained within the transponder noise bandwidth, defined by its bandpass response. The total noise power N_S measured at some reference point in the receiver is given by

$$N_S = N_A + N_R, \text{ (watts)} \tag{27-9}$$

where the subscript S indicates "system," which in this case refers to the uplink receiving system. Additional noise is added on the downlink, and we shall discuss later how the video signal-to-noise ratio for the overall link is affected by both uplink and downlink noise.

It is convenient to measure noise power in terms of an *effective noise temperature*. Any noise power NP_x is given by

$$NP_x = k T_x B_{NP}, \text{ (watts)} \tag{27-10}$$

where

T_x is the effective noise temperature at the reference point in Kelvin (K)

K = Boltzmann's constant ($1.38 \cdot 10^{-23}$ Joule/K)

B_N is the noise bandwidth in hertz, determined here by the transponder bandwidth

Thus

$$T_x = \frac{NP_x}{k B_{NP}} \tag{27-10a}$$

Division of Eq. (27-9) by $k B_N$ gives

$$TS = T_A + T_R \text{ (K)} \tag{27-11}$$

T_S is called the operating temperature or the system temperature, where S for "system" again refers to the uplink receiving system.

The carrier-to-noise power ratio C/N_S is fundamental to evaluating link performance. It is obtained by dividing Eq. (27-8) by Eq. (27-10), giving

$$C/N_S = \frac{GR}{T_S} \left[\frac{EIRP \, k_A}{k B_{NP}} \left(\frac{\lambda}{4\pi R} \right)^2 p \right] \tag{27-12}$$

Examination of Eq. (27-12) shows that C/N_S is proportional to G_R/T_S. This ratio, usually designated simply G/T, is at least to some extent under the control of the satellite designer, as opposed to the factors within the brackets. Thus, the ratio G/T, usually expressed in dB, is called the *figure of merit* of the satellite receiving system.

It is convenient to express C/N_S in decibels for ease of calculation. First, from Eq. (27-8):

$$C = EIRP - L_A + G_R - L_S - L_p, \text{(dBW)} \tag{27-13}$$

where $EIRP$ is expressed in dBW, and where L_A, L_P, and L_S (L_S is usually called "space loss," or more appropriately "spreading loss") are all expressed in dB, respectively by

$$L_A = -10 \log k_A \tag{27-13a}$$

$$L_p = -10 \log p \tag{27-13b}$$

$$L_S = 20 \log (4\pi R / \lambda) \tag{27-13c}$$

If we write Eq. (27-10) in decibel form and substitute the numerical value of k, we obtain

$$NPS \text{ (dBW)} = T_S(\text{dB}/K] + 10 \log B_N - 168.6 \tag{27-14}$$

where B_{NP} is expressed in MHz.

From Eqs. (27-13) and (27-14) we can write, dropping the subscript S in writing C/N,

$$C/N \text{ (dB)} = EIRP - L_A - L_S - L_p + G/T \text{ (dB/K)} - 10 \log B_{NP} + 168.6$$

where

$$C/N \text{ (dB)} = C(\text{dBW}) - NS \text{ (dGW)} \tag{27-15a}$$

and

$$G/T \text{ (dB/K)} = G_R(\text{dB}) - T_S \text{ (dB-K)} \tag{27-15b}$$

In Fig. (27-7), the counters of saturation flux density coincide with the counters of $G/T(\text{dB/K})$, where G is the gain of the satellite receiving antenna and T is the effective system operating noise temperature, both measured at the same point in the satellite receiving system.

For our case, reading $G/T = -4$ dB for the signal path to the satellite from New York City. Assuming L_p and $L_A = 0$, from Eq. (27-15) we have for the assumed 36 MHz transponder noise bandwidth.

$$C/N \text{ (dB)} = 80 - 199.6 - 4 - 15.6 + 168.6 = 29.4 \text{ (dB)} \tag{27-16}$$

It will be seen that this value of uplink C/N contributed only a small fraction of the total noise of a typical communications link, and that the downlink usually contributes most of the noise. This is to be expected in a well designed uplink, because, if the uplink is noisy, it is not possible to realize a high quality communications link regardless of the quality of the downlink.

27.3 Downlink

To begin we shall discuss the satellite's *traveling wave tube* amplifier (TWT, or TWTA). As shown in Fig. 27.6, the receiver feeds a down-converter, which offsets the frequency and drives the downlink TWT output power amplifier. In keeping with our simple assumptions, the frequency is shown as offset by 2 GHz instead of 2.225 GHz.

The transponder TWT maximum output is relatively low, typically between 5 and 10 watts. Figure 27.8 shows a typical TWT power input-output curve.

When only one carrier exists, as we assume in our case, the TWT is usually operated under *saturated* conditions. When more than one carrier is present, as is often the case in real-world applications, the TWT must be operated in its linear region to control intermodulation distortion, requiring a back-off in input power.

The satellite output is measured by the *EIRP* defined by Eqs. (27-1) or (27-2). The satellite *EIRP* is lower than that of the uplink earth station because of two factors. First, the power obtained from the solar panels which supply the

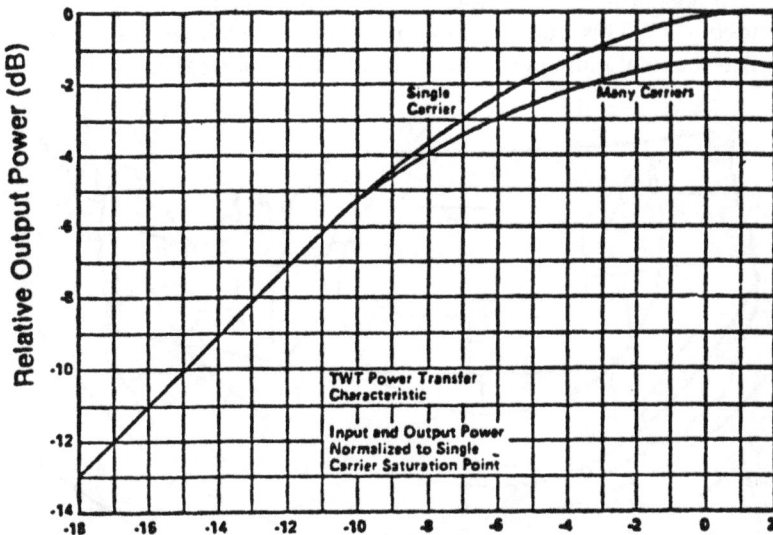

Figure 27.8 TWT Power-Transfer Characteristic Showing Linear and Saturation Regions
Source: Scientific Atlanta

satellites is limited by satellite size and launch costs. Second, downlink coverage is provided over a wide area (usually about the same as the uplink G/T footprint), and this limits the gain of the satellite transmitting antenna. A typical downlink footprint, described by contours of constant *EIRP*, is shown in Fig. 27.9.

The operation of the downlink is basically similar to that of the uplink, except for the lower *EIRP* produced by the satellite. The power density at the receiving antenna is given by Eq. (27-3), where R is now the distance from the satellite to the receiving earth station, and k_A is the atmospheric attenuation factor for this path at the downlink frequency.

A simplified block diagram of the receiving earth station is shown in Fig. 27.10. It consists of the earth station antenna, a low noise amplifier (LNA), receiver, and associated equipment.

As in the case of the satellite, the receiving earth station is characterized by a figure of merit, G/T, whose required value is determined by the original video signal-to-noise ratio of 54 dB. We shall derive this required value below.

The downlink carrier-to-noise ratio is again given by Eq. (27-15), where the various terms represent downlink parameters.

In describing the operation of the uplink, it is convenient to write Eq. (27-15) in terms of C/N. This is done because it is easy to visualize the ratio of carrier power to the total noise power in the transponder passband.

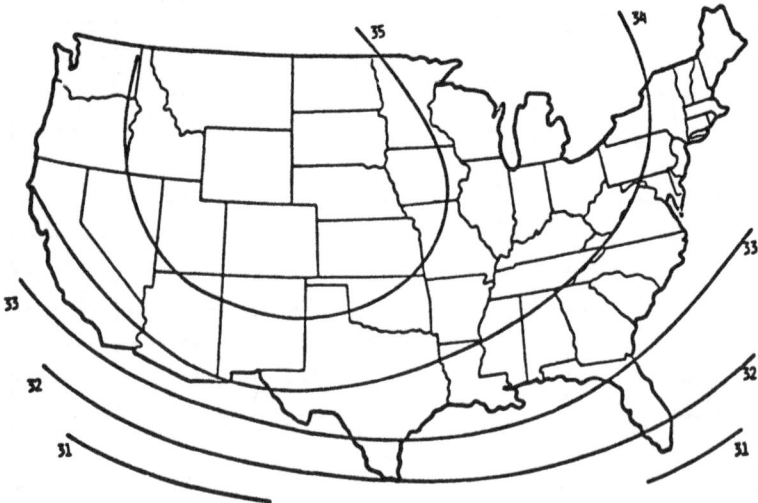

Figure 27.9 Footprint Showing Assumed Satellite EIRP in dBW at 4 GHz
Source: Scientific Atlanta

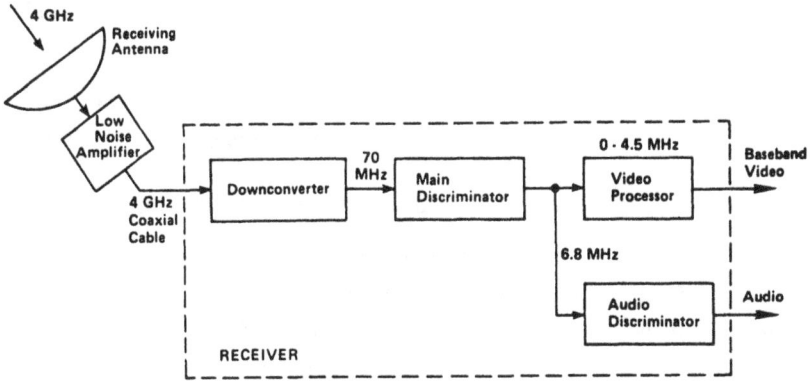

Figure 27.10 Simplified Block Diagram of Assumed Satellite Receiving Earth Station. Note: The LNA is bolted directly to the output waveguide of the receiving antenna to eliminate losses which would otherwise increase the system noise temperature.

Source: Scientific Atlanta

However, in analyzing a link, it is generally more meaningful to calculate the carrier-to-noise/power-density ratio, C/N_o, where N_o is the noise power density, usually expressed in watts/MHz; from Eq. (27-10) N_o is given by $N_o = kT(10^6)$. The relationship between C/N_o and C/N is given by

$$C/N_o \text{ (dB – MHz)} = C/N_o \text{ (dB)} + 10 \log B_N \text{ (MHz)} \qquad (27\text{-}17)$$

where B_N is the IF bandwidth of the earth station receiver.

C/N_o is used because the video signal-to-noise ratio for a given deviation of the video signal is determined by C/N_o and the noise bandwidth of the baseband filter function of the receiver, and not by the predetection carrier-to-noise power ratio C/N. For a US television signal, with CCIR weighting, and with peak deviation of 10.75 MHz, the video signal-to-noise ratio is given by

$$S/N = C/N_o + 22.6, \text{ (dB)} \qquad (27\text{-}18)$$

For the specified video signal-to-noise ratio of 54 dB, the required overall link C/N_o is thus 31.4 dB/MHz. This is actually the required overall link C/N_o, but we will use this value for the downlink and show that the required downlink $C/N_o = 31.6$ dB (the uplink noise is taken into account). A margin in uplink power is provided to account for noise and other factors.

Rewriting Eq. (27-15) in terms of C/N_o by use of Eq. (27-17) and rearranging give

$$G/T = C/N_o - EIRP + L_S - 168.6, \text{ (dB/K)} \qquad (27\text{-}19)$$

where L_A and L_p are assumed to be zero.

For example, we will read an *EIRP* of 33.5 dBW for Los Angeles from Fig. 27.9. For the distance from Los Angeles to the satellite, L_S is found to be 196 dB from Eq. (27-13c).

Substitution of the appropriate parameters in Eq. (27-19) gives

$$G/T = 31.4 - 33.5 + 196 - 168.6 = 25.3, \text{ (dB/K)} \qquad (27\text{-}20)$$

This is the figure of merit of the earth station receiving system, which is required to provide a downlink C/N_o of 31.4 dB-MHz under the assumed conditions.

Next, let us consider the receiving system noise temperature. As in the case of the uplink, T_S is given by Eq. (27-11). The antenna temperature T_A of the earth station can be made very low because the earth station beam looks at the cold sky. T_A for the chosen antenna varies from 46°K at an elevation angle of 5° to 18°K at zenith. At an elevation angle of 34°, it is approximately 20°K.

In Eq. (27-11) T_R is almost equal to T_{LNA}; i.e., the temperature of the low noise amplifier for a well designed earth station.

From the definition of G/T given in Eq. (27-15b), the required G/T given by Eq. (27-20) is obtained by subtracting the receiving system noise temperature from the receiving antenna gain. Thus, we can trade off one against the other. This is illustrated by the graph of Fig. 27.11.

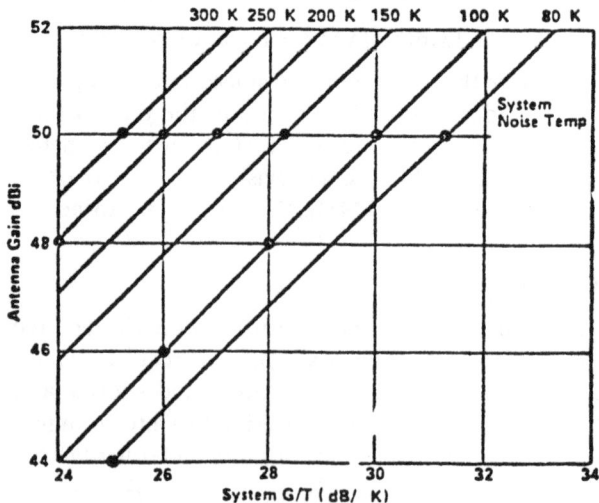

Figure 27.11 System G/T vs. Antenna Gain
Source: Scientific Atlanta

The Scientific-Atlanta model 8010 7-meter antenna, shown in Fig. 27.12, has a gain of 47.5 dB at 4 GHz. We will choose this antenna, which results in a required maximum system noise temperature T_s of 22.2 dB/I for the G/T of 25.3 dB/K. This T_s translates to $166°$ [= 10 exp(22.2/10)]. We will find that this required system noise temperature is easily attained at moderate cost.

On the other hand, if we were to choose a 5-meter antenna with a gain of 44.5 dB, the resulting required system noise temperature would be 19.2 dB/K or $83°$. This noise temperature is not realizable with an uncooled LNA.

The best choice is the 7-meter antenna and a moderate cost $100°$K LNA.

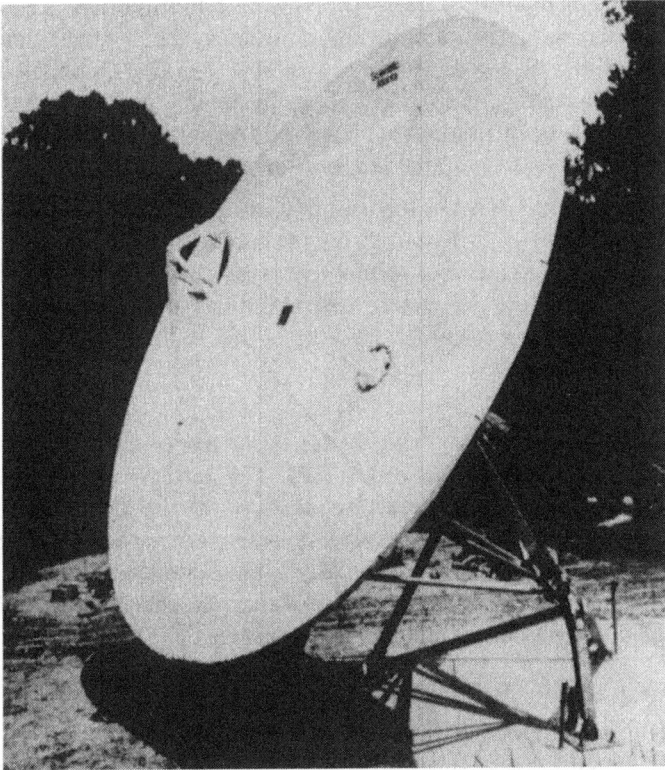

Figure 27.12 Scientific-Atlanta Model 8010 7-Meter Satellite Communications Antenna
Source: Scientific Atlanta

Taking into account the antenna noise temperature of 20°K, the downlink system noise temperature is 120°K, or 20.8 dB/K. From Eq. (27-15b), the resulting G/T is given by

$$G/T = 47.5 - 20.8 = 26.7, (dB/K) \tag{27-21}$$

This represents a margin of 1.4 dB over the value of 25.3 dB/K given by Eq. (27-20).

27.4 LNA

Low noise amplifiers (LNAs) used in receiving earth stations (Fig. 27.10) are Gallium-Arsenide field effect transistor (GaAs FET) amplifiers. Research in recent years has reduced the noise figure of GaAs FETs to the point that uncooled amplifiers as low as 80°K are available. The LNA is botled directly to the output waveguide flange of the receiving antenna, where T_A is defined, to eliminate noise contributions which would be introduced by waveguide losses if the LNA were separated from the antenna.

In our example, the LNA has a gain of 50 dB. The receiver is located in a convenient position, some distance from the antenna, and is connected to the LNA output by a coaxial cable, whose loss is assumed to be 10 dB. The high gain of the LNA almost completely overrides losses and noise contributions arising in the cable and circuitry.

27.5 Receiver

The receiver, shown in Fig. 27.10, converts the frequency of the incoming signal to an intermediate frequency (IF). The receiver rejects unwanted signals, demodulates the IF signals by means of the main IF discriminator, and processes the composite baseband video to restore it to its pre-transmission format. The audio subcarrier is separated from the main IF and is demodulated by the subcarrier discriminator. Baseband video and audio are available at separate output coaxial connectors.

27.6 Overall Link C/N$_o$ and C/N

From Eq. (27-18), the required value of overall link C/N_o was 31.4 dB/MHz. For convenience, we used this value for the downlink C/N_o. Because of the contribution of the uplink noise to the overall link noise, the required downlink noise is actually slightly greater than 31.4 dB/MHz, giving an overall link C/N_o of 31.4 dB/MHz.

The overall link C/N_o results from adding the noise power densities of the uplink and downlink referred to the same carrier levels:

$$(N_o/C)_o = (N_o/C)_U + (N_o/C)_D \tag{27-22}$$

where the subscripts o, U, and D represent "overall link," "uplink," and "downlink," respectively, and where the terms in parentheses are numerical ratios (not decibels).

Thus, for given overall and uplink carrier-to-noise power densities,

in our example

$$(N_o/C)_D = (N_o/C)_U + (N_o/C)_D \tag{27-23}$$

and

$$(N_o/C)_o = \frac{1}{10^{31.4/10}} = \frac{1}{1380} \tag{27-24}$$

$$(N_o/C)_U = \frac{1}{10^{45/10}} = \frac{1}{31623} \tag{27-25}$$

From Eq. (27-23)

$$(C/N_o)_D = \frac{1}{(N_o/C)_D} = \frac{1}{\dfrac{1}{1380} - \dfrac{1}{31623}} = 1443 \tag{27-26}$$

The required downlink carrier-to-noise power density C/N_o is thus given by

$$C/N_o = 31.6 \text{ (dB/MHz)} \tag{27-27}$$

Note that the additional 0.2 dB over the originally indicated 31.4 dB/MHz is included in the 1.4 dB margin.

The final system C/N is determined from Eq. (27-17), where B_N is the earth station receiver IF bandwidth (except in the unlikely event that the satellite transponder bandwidth is narrower than the receiver bandwidth and thereby determines the system bandwidth).

27.7 Link Performance

At the beginning of this chapter, we set a specification of 54 dB for the video signal-to-noise ratio, S/N, and indicated that *distortion* limits would be discussed.

The S/N criterion is satisfied by the carrier-to-noise power density ratio of the overall link for the FM deviation used, as indicated by Eq. (27-18). Linear and nonlinear distortions of the output wave form are introduced by variations in the amplitude and group-delay responses of the transmitter, satellite, and receiver. To maintain distortion within acceptable bounds, specifications have been established which limit group-delay and amplitude responses of the receiver and the transmitter. These distortion specification limits are defined by a mask within which the responses must fall. Examples of such masks are shown in Fig. 27.13.

Turning to the distortion limits for our simplified example, after the complete link has been installed, the uplink amplitude and group-delay responses are adjusted to fall within the limits set by the masks. If the receiver has been compensated separately such that its response falls within the same mask, the total variation in group-delay response will be approximately twice that defined by the mask.

In addition to RF tests, real-time baseband measurements are made as in any standard video link to test total input-to-output link distortion.

27.8 Threshold

In satellite FM-video systems, as described in our simplified example, a high deviation (10.7 MHz), and hence a wide occupied bandwidth (approximately 36 MHz), are used to increase video signal-to-noise ratio S/N over the IF carrier-to-noise power density ratio C/N_o.

We previously discussed noise in relation to the value of C/N_o required to produce a desired value of S/N. This enhancement in S/N occurs only if C/N is above a certain threshold level. The effect of this threshold level is shown by Fig. 27.14, which is a graph of S/N as a function of C/N of a typical video receiver.

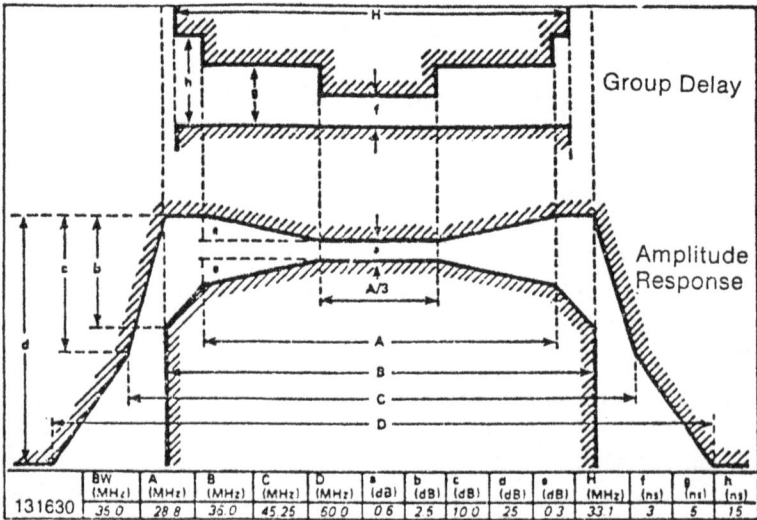

131630	BW (MHz)	A (MHz)	B (MHz)	C (MHz)	D (MHz)	a (dB)	b (dB)	c (dB)	d (dB)	e (dB)	H (MHz)	f (ns)	g (ns)	h (ns)
	35.0	28.8	36.0	45.25	50.0	0.6	2.5	10.0	2.5	0.3	33.1	3	5	15

Figure 27.13 INTELSAT 36 MHz Group-Delay and Amplitude-Response Masks

Source: Scientific Atlanta

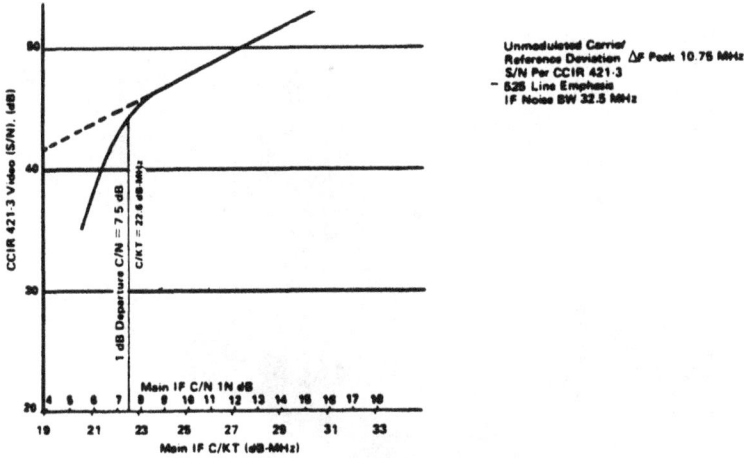

Figure 27.14 Illustration of FM/Video Receiver Threshold
Source: Scientific Atlanta

When the C/N ratio at the discriminator is above the threshold, there is a one-to-one ratio between video signal-to-noise power and IF carrier-to-noise power. As the IF C/N ratio passes through the threshold, random noise peaks begin to cause the instantaneous envelope of the carrier-plus-noise voltage to pass through zero. When this happens, sudden phase changes occur in the IF signal, which the discriminator interprets as being caused by large instantaneous frequency changes. This results in impulse noise spikes in the baseband signal, causing white-to-black and black-to-white streaks on the video display. The rate of occurrence increases as C/N decreases, and, as C/N decreases still further, loss of synchronization occurs and the picture is lost.

Other satellite communications systems, such as digital links, might not exhibit pronounced thresholds, depending on the demodulation process.

27.9 Satellites

The simplified satellite described in our example is assumed to contain a single transponder. In practice, however, operational 6/4 GHz satellites in orbit over the US have either 12 or 24 transponders. These communications satellites, of course, contain block converters, diplexers, and other circuits not shown in the simple satellite example of Fig. 27.6.

Figure 27.15 is a photograph of the RCA SATCOM satellite, as an example. Figure 27.16 is the transmission and reception frequency plan of its 24 transponders. The COMSTAR satellite also has 24 transponders, made

Figure 27.15 Satellite Photograph
Source: Scientific Atlanta

possible by the technique of frequency re-use, while the older generation WESTAR and ANIK satellites employ only 12 transponders.*

If left alone, a geosynchronous satellite would finally drift out of its orbital position. To avoid this, it is continuously monitored by a *telemetry tracking and control* (TT&C) station. Small jets of a propellant such as hydrazine are used to keep it within a "station-keeping box," and sufficient propellant is carried on-board the satellite for its predicted life, which is usually in the range of 7 to 10 years. Station-keeping boxes for 6/4 GHz satellites serving the US are ±0.1° on each side. This amount of drift is small enough that antennas which are less than about 10 or 11 meters in diameter can usually be left stationary. Larger antennas may contain a sensing mechanism to keep the antenna beam peaked on the satellite.

The solar panels, which power the satellite by converting sunlight directly to electricity, must face the sun in order to be effective. The *attitude* of the satellite also must be stabilized to keep its antenna pointing in the desired direction.

Some satellites are designed so that the solar panels are continuously pointed toward the sun, with their antennas independently pointed toward the earth,

*See D. Jansky, ed., *World Atlas of Satellites* (Dedham, MA: Artech House, 1983).

as in the case of SATCOM and INTELSAT V. Many satellites — including INTELSATs I through IV A, WESTAR, ANIK, and COMSTAR — are cylindrical in shape, and the solar cells are mounted around the periphery. The bodies of these satellites spin about their axes for stabilization, with the antennas always pointing toward the earth. In this way, approximately one-third of the cells effectively face the sun at a given time. Batteries are used in almost all satellites to take care of solar panel outages during eclipses of the satellite by the earth.

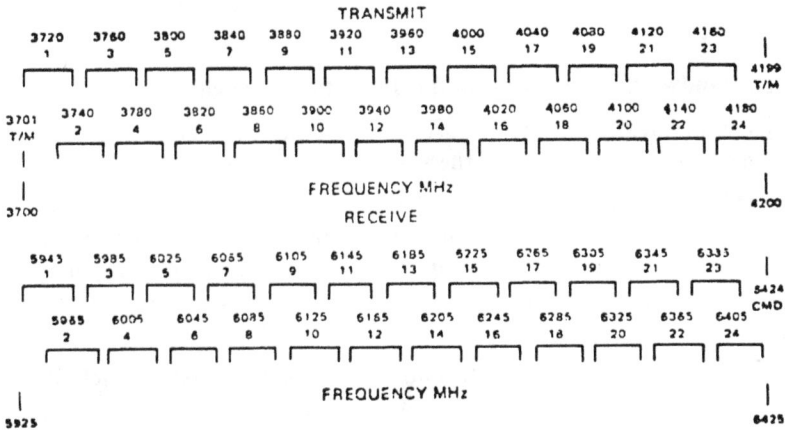

Figure 27.16 Transmit-Receive Frequency Plan of RCA Satcom Satellite
Source: Scientific Atlanta

27.10 Frequency Re-Use

Consideration of the RCA SATCOM frequency plan of Fig. 27.16 shows that the odd-channel sidebands overlap those of the even channels. This overlap occurs because the newer satellites are designed to transmit and receive signals simultaneously on two orthogonal polarizations, increasing the number of transponders in the available 500 MHz band from 12 to 24.

The polarization of all of the US and Canadian domestic satellites are linear. Linear polarizations which are at right angles to each other are electrically *orthogonal*. If the polarization of the receiving antenna is orthogonal to that of the incident wave, the polarization efficiency p in Eq. (27-6) is zero, and no power is received from the wave. The orthogonal polarization can then be used to receive independent information.

Frequency re-use is accomplished by staggering the carriers of the odd and even channels so that only sideband energy overlaps. The carriers are staggered because it would be too difficult to design systems which reject the unwanted signal by polarization discrimination alone. Staggering the carriers reduces the required system polarization discrimination from >50 dB to about 25 dB, a value which is achievable unless exceptionally hard rains cause additional depolarization. Even under conditions of severe depolarization, which occur only occasionally, the interference to the orthogonal channel is usually not extreme because of the protection afforded by the staggered carrier-frequency.

27.11 Bandwidth, Data Rate, and Information Capacity

Any communications system must carry out the basic task of transmitting information from the input of the system to the output.

Transfer of information implies changes in signal level *versus* time. The *information rate*, as the term implies, is the time rate of information transfer. (Information rate is also called *data rate*.) The information capacity of a system is determined by the maximum rate at which it can transmit information.

System information capacity depends on a number of factors, including the system bandwidth, the power used, the sensitivity of the receiving equipment (i.e., suppression of noise), and the modulation and demodulation techniques employed.

The system information capacity is not determined by the amount of information to be transferred, but by the rate at which it can be transferred. As examples:

- Pictures with extremely fine detail can be transmitted by the spacecraft VOYAGER from Jupiter or Saturn by a relatively low information-capacity system by taking a long period of time to transmit each picture.
- Transmission of a television signal in real time requires a relatively high information-capacity system because the signal contains a large number of level changes, representing brightness and color, which must be transmitted at a faster rate.
- The information capacity required to transmit a voice signal in real time is much less than that required to transmit a television signal because the required rate of information transfer is much lower.

The information capacity of a system cannot be increased indefinitely. Energy is required to change a voltage from one level to another, and an infinite information rate implies an infinite rate of change of energy, which is not possible. In addition, an infinite amount of information implies subdividing levels into infinitesimal increments. This cannot be done because every system contains noise which obscures the detail.

Therefore, for any system there is a minimum time τ required for a signal to change from one level to another, and a maximum number n of level changes that it is designed to detect.

The information content I of a message which is T units of time in length can be expressed by

$$I = \frac{T}{\tau} \log_2 n, \text{ (bits)} \tag{27-28}$$

where the *bit* is a unit of information. Thus, information capacity is defined quantitatively as *the number of bits per second that can be transferred.*

An *analog* system is one in which the output is a continuous replica (i.e., analogy) of the input signal except for distortion and noise. In an analog system, n in Eq. (27-28) is effectively large. In *digital* systems, n is a small number sensed numerically by a code to represent the signal. In a *binary* system, n is 2, and only two levels, usually labeled "zero" and "one," are sensed.

Video and noise signals are generally transmitted by analog, while computer data, for example, are transmitted by digital. *Analog-to-digital converters* (A/D) are often used to transmit analog signals via digital systems. The digital signals are then reconverted to analog form at the receiving end of the system by *digital-to-analog converters* (D/A). Such systems are increasingly used as the cost of digital circuits decreases.

Electrical wave forms can be represented in either the *time domain* or the *frequency domain*, and there is a direct relationship between the data rate and the minimum bandwidth required to transmit information.

Although the bandwidth occupied by complicated, time-varying signals cannot be specified exactly, the bandwidth B_s which contains most of the energy can be related to the width of Eq. (27-21) by

$$B_S = \frac{k}{\tau} \tag{27-29}$$

where k is a constant whose order is unity and whose specific value depends on the criterion selected.

Virtually all satellite communications systems use some form of *angle modulation* (for example, FM). To generate angle modulation, the phase or the frequency (which is proportional to the time derivative of the phase), is varied linearly with the modulating signal.

The *occupied bandwidth* for FM is determined by both the frequency deviation of the carrier and by the modulating signal frequencies. Specifically, the FM bandwidth (i.e., that occupied by sidebands of significant level) is given by Carson's rule:

$$B_C = 2 (f_m + \Delta f) \qquad\qquad\qquad (27\text{-}30)$$

where

B_C = *Carson's-rule bandwidth*
f_m = *highest modulation frequency of interest*
Δf = *peak deviation of the carrier*

In domestic satellite FM-video systems a high deviation (10.7 MHz), and hence a wide occupied bandwidth (approximately 36 MHz), are used to gain a significant increase in video signal-to-noise ratio over the IF carrier-to-noise density ratio C/N_o.

In digital systems, modulation formats are usually used which give a low bit error rate for a given ratio E_b/N_o, where E_b is the energy per bit.

Existing satellites and associated earth-station equipment are designed to cover a wide range of data rates, from television signals or high-rate streams of digital data to low data rates as represented by individual audio channels.

27.12 Modulation Formats and Access Techniques

The modulation techniques employed for various satellite communications applications depend on the particular requirements of the application.

Where many narrow-band channels, such as voice-grade circuits, are required, the narrow-band signals are *frequency-division multiplexed* (FDM) and are used to frequency modulate a carrier. This modulation method is called FDM/FM.

For full-transponder video, or multiple-channel voice-grade circuits, each transponder is accessed by a single earth station at a given time. For lower capacity voice, data, and half-transponder video applications, a transponder will be shared by a number (n) of earth stations, each of which uses a number (m) different carrier frequencies. This technique is called *frequency-division multiple access* (FDMA).

In one of the important applications of FDMA a single voice-grade channel is transmitted on each carrier. This approach, designated *single-channel per carrier* (SCPC), uses carriers spaced 60 kHz or less apart.

Although SCPC requires a high degree of frequency stability from the individual carriers, it permits installation of stations which need only a limited number of voice-grade channels and allows for easy addition of more channels as required.

SCPC with *demand assignment multiple access* (DAMA) is a technique in which each earth station uses a channel only as required. When the earth station is not using a channel, it is available for use by other earth stations. DAMA greatly increases the average number of voice channels that can be carried by one transponder. DAMA is in operation on the INTELSAT SPADE system and the Marisat system, and it is finding increasing application in domestic and foreign thin-route systems.

Biphase and *quadriphase* digital modulation (PSK and QPSK) formats are becoming increasingly important as digital technology advances and the cost of digital circuitry decreases.

Systems with *time-division multiplex multiple-access* (TDMA) and TDMA with *satellite switching* (SSTDMA) are also in use or under development. Satellite Business Systems, (a partnership among wholly-owned subsidiaries of Comsat General, IBM, and Aetna Life and Casualty), is implementing an extensive TDMA multipoint communications system using a single, time-shared carrier per transponder with transmission using QPSK in a burst mode. This system will operate in the 14/12 Ghz band and can be located directly on customers' premises.

Western Union is including a digital format with satellite switching, operating in the 14/12 Ghz band on its ADVANCED WESTAR satellite, which is part of a shared space segment with the NASA tracking and data relay service system (TDRSS). The ADVANCED WESTAR segment will use high gain spot-beams in a switching matrix to provide interconnection of six antenna beams in an SSTDMA system. The high gain provided by the spot-beams helps overcome problems caused by rain attenuation at 12 and 14 GHz.

These and various other techniques are utilized to take advantage of the available spectrum and provide system flexibility to accommodate the various categories of users and applications.

27.13 Effect of Frequency on Satellite Communications Systems

Because of the virtually unlimited applications for satellite communications and the limited information capacity of the 6/4 GHz band, higher frequency satellite bands, especially the 14/12 GHz band, will come into greater use in the future. Therefore, it is interesting and important to consider the effect of increasing frequency on the design and performance of satellite communications systems.

The major advantages of using higher frequencies are (1) the availability of more spectrum, (2) the fact that less interference exists from terrestrial microwave systems, and (3) the production of higher gain antennas of a given aperture size.

The negative factors involved in using the higher frequencies are (1) the increase in atmospheric attenuation with frequency (especially that due to rain), (2) the decrease in sensitivity (i.e., increase in noise temperature) of receivers, (3) the narrower beamwidths produced by antennas of a given aperture size, and (4) the smaller surface tolerances required to prevent degradation of antenna performance at the shorter wavelengths of the higher frequencies.

Let us consider the effect of frequency on an earth-station receiving antenna. Remember, the gain of an antenna of a given diameter increases as the square of the frequency. This increase in gain does not mean that an antenna of a given size receives more power for a given satellite *EIRP* as the frequency increases. This is because the received power, a given by Eq. (27-6), is independent of frequency if the effective area of the antenna is constant. Therefore, the realizable effective area is essentially independent of frequency for reflector-type antennas of a fixed size.

Thus, as frequency increases, the major effect is a narrowing of the beam and a decrease in surface tolerance of the reflector. For a given satellite *EIRP*, the increase in frequency increases the earth-station receiving antenna cost in at least three ways:

1. By *tighter surface tolerance*, which requires a more accurate and stiffer reflector;

2. By the increased *difficulty of pointing the antenna* toward the satellite, which requires a more costly antenna mounting structure; and

3. By the fact that the increase in atmospheric attenuation and the inherent increase in LNA noise temperature dictate a *larger antenna diameter*, which acts to reinforce the difficulties associated with tighter surface tolerance and antenna pointing difficulty.

In sum, if earth station costs are to be kept low, the *EIRP* of the satellite must be greater at the higher frequency band so that smaller antennas can be used. This increases the cost of the satellite because the increase in *EIRP* must come from either higher power transponders or footprints covering smaller areas.

Obviously, the smaller footprints do not represent a disadvantage where a small area is to be covered such as Japan or a country in Europe. On the other hand, where a large area is to be covered, such as the United States, smaller footprints require multiple beams, each with some number of transponders determined by the level of service to be provided. The result is an increase in satellite solar panel power requirements. Typical system designs for low cost receiving systems are based on four or five beams covering the continental, or contiguous, US (referred to as CONUS).

Despite the problems, movement toward higher frequencies for new satellites is inevitable for the US because of orbital crowding and use of the 6/4 GHz band by both terrestrial and satellite systems. Systems are already being implemented for the 14/12 GHz band. In fact, once sufficient cost is transferred to the satellite so that small antennas can be used and sufficient system information capacity is provided, many new applications open up such as direct broadcast (DBS). However, satellite systems at 6/4 GHz are here to stay because of their inherently low cost and the existing frequency spectrum.

28 Earth Station Antennas

An *antenna* is a device for accomplishing a transition between a guided electromagnetic wave and a wave propagating in free space. The earth station antenna is the connecting link between free space and the receiver, and plays a vital role in determining system performance.*

28.1 Types of Earth Station Antennas

Several types of earth station antennas are now in use within the US and abroad. These antennas can be grouped into two broad categories: *single beam antennas* and *multiple beam antennas*. A single beam earth-station antenna is defined as an antenna that generates a single beam which is pointed toward a satellite by means of a positioning system. A multiple beam earth-station antenna is defined as an antenna which generates multiple beams by employing a common reflector aperture with multiple feeds illuminating that aperture. The axes of the beams are determined by the location of the feeds. The individual beam identified with a feed is pointed toward a satellite by positioning the feed without moving the reflector. The majority of the earth station antennas now in use are single beam antennas.

Single beam antenna types used as earth stations are paraboloidal reflectors with focal-point feeds (prime-focus antenna); dual-reflector antennas, such as the Cassegrain and Gregorian configurations; horn-reflector antennas; offset-fed paraboloidal antennas; and offset-fed, multiple reflector antennas. Each of these antenna types has unique characteristics, and their advantages and disadvantages must be considered when choosing them for a particular application.

---

*See L.V. Blake, *Antennas*, 2nd ed., (Dedham, MA: Artech House, 1984).

28.2 Axisymmetric Dual-Reflector Antennas

The predominant choice of most system operators has been the *dual-reflector Cassegrain antenna*. Cassegrain antennas can be divided into three primary types:

Type 1. The classical Cassegrain geometry employing a paraboloidal contour for the main reflector and a hyperboloidal contour for the subreflector (see Fig. 28.1). The paraboloidal reflector is a point focus device with a diameter D_p and a focal length f_p. The hyperboloidal subreflector has two foci. For proper operation, one of the two foci is the real focal point of the system, and is located coincident with the phase center of the feed. The other focus, the virtual focal point, is located coincident with the focal point of the main reflector.

Type 2. A geometry consisting of a paraboloidal main reflector and a special shaped, quasi-hyperboloidal sub-reflector. The geometry in Fig. 28.1 is appropriate for describing this antenna. The main difference between this and the classical Cassegrain mentioned above is that the sub-reflector has been designed such that the overall efficiency of the antenna has been enhanced,

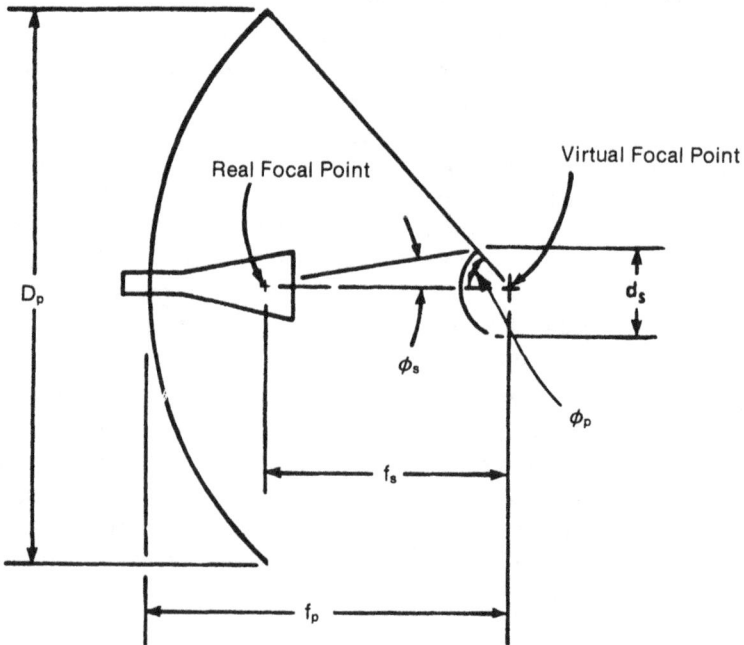

Figure 28.1 Geometry of the Cassegrain Antenna System
Source: Scientific Atlanta

thereby yielding improved gain performance. This technique is especially useful with antenna diameters of approximately 30 to 100 wavelengths; for example, a 5-meter antenna in the 6/4 GHz frequency band.

Type 3. Research has shown that in dual reflector systems with high magnification — essentially, a large ratio of main reflector diameter to sub-reflector diameter — the distribution of energy (as a function of θ) is largely controlled by the sub-reflector curvature. The path length, or phase front, is dominated by the main reflector (see Fig. 28.2). A method was found to simultaneously solve for the main reflector and sub-reflector shapes to obtain an exact solution for both the phase and amplitude distributions in the aperture of the main reflector of an axisymmetric dual-reflector antenna. This technique, based on geometrical optics, was highly mathematical and involved solving two simultaneous, nonlinear, first-order, ordinary differential equations. These methods have been applied to design axisymmetric dual-reflector antennas when maximum gain is needed for a given size reflector antenna. The method of solution allows the specification of an arbitrary amplitude distribution, whereby one can make compromises between sidelobes and antenna efficiency (see Fig. 28.2).

28.3 Prime-Focus-Fed Paraboloidal Antenna

The other type of antenna most often employed in the US for receive-only applications is the *prime-focus-fed paraboloidal* (see Fig. 28.3) reflector. For large aperture sizes, this type of antenna has excellent sidelobe performance in all angular regions, except the "spill-over" region around the edge of the reflector. Even in this area, sidelobe suppression can be achieved, which satisfies the FCC pattern requirement. Antenna efficiency for apertures greater than 100 wavelengths is about 60%. Therefore, the prime-focus-fed

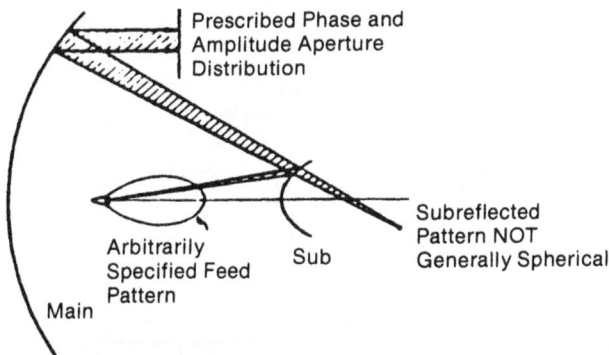

Figure 28.2 Circularly Symmetric Dual Shapped Reflectors
Source: Scientific Atlanta

reflector represents a good compromise choice between sidelobe and gain performance factors. For aperture sizes less than approximately 40 wavelengths, the blockage of the feed and feed-support structure raises the sidelobes with respect to the peak of the main beam, such that it becomes exceedingly difficult to meet the FCC sidelobe specification.

28.4 Horn-Reflector Antenna

Two other types of single beam antennas are used in earth stations. These are the *horn reflector* and *offset-fed reflector*.

The *pyramidal horn-reflector* has been used for many years in the terrestrial microwave field, primarily by AT&T. This antenna offers excellent efficiency and sidelobe performance, especially in the near region and the back region.

The *conical horn-reflector* is similar in design to the rectangular horn-reflector. The first sidelobe is improved by its circular aperture. The conical horn has less wind resistance than the pyramidal horn, again as a result of its basic shape. Both the conical and pyramidal horns have unique mounting requirements which are somewhat restrictive and cumbersome. For example, the overall length of the antenna is typically twice the aperture size. (See Fig. 28.4 for the dimensions of a typical 4-meter conical horn-reflector.) Transportation problems can also occur because of the large physical dimensions of the antenna.

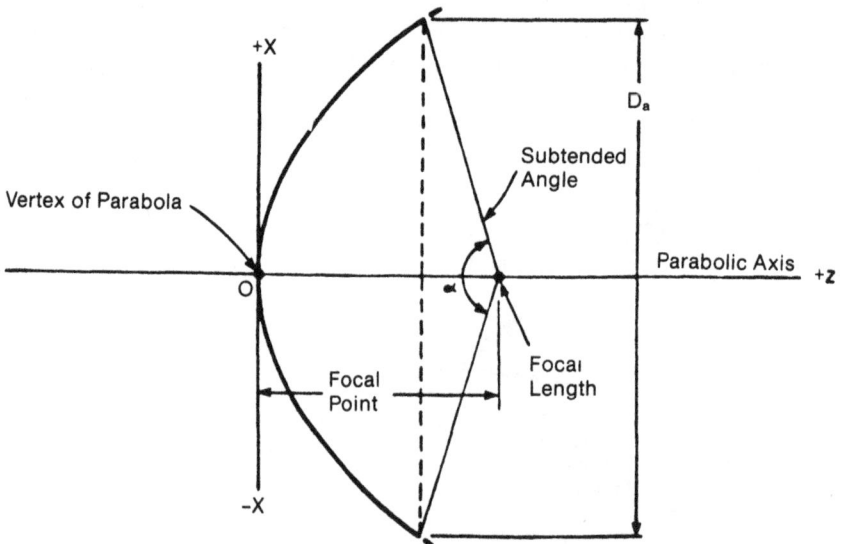

Figure 28.3 Geometry of Prime Focus Antenna
Source: Scientific Atlanta

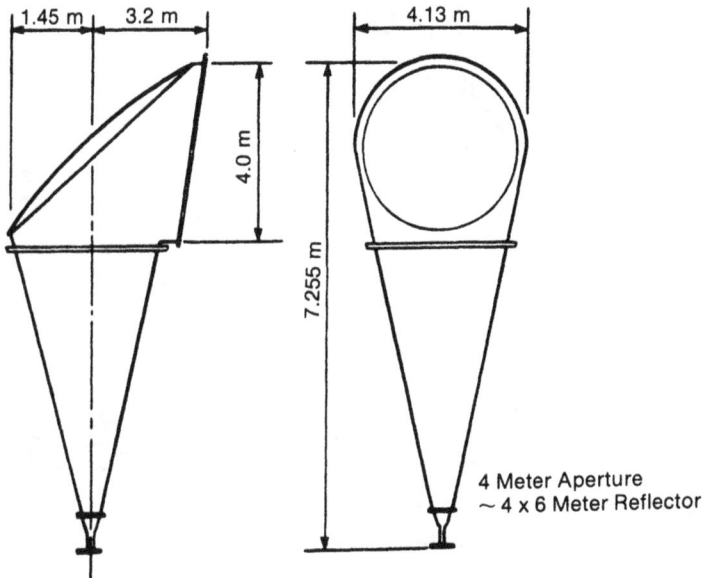

Figure 28.4 4-Meter Conical Horn Reflector Antenna
Source: Scientific Atlanta

28.5 Offset-Fed Reflector Antennas

Generally, the *offset-fed reflector* antenna has not been widely used as an earth station antenna because of its cost. This probably will remain the case for the near future. The offset-fed reflector antenna can employ a single main reflector, or multiple reflectors, with dual-reflectors the more prevalent of the multiple reflector designs. The offset, front-fed reflector, consisting of a section of a paraboloidal surface (Fig. 28.7), minimizes diffraction scattering by eliminating the aperture blockage of the feed, and feed-support structure. Sidelobe levels of (320-330) log θ dBi and aperture efficiencies of 60 to 70% can be expected from this type of antenna. The increase in aperture efficiency as compared to an axisymmetric prime-focus-fed antenna is a result of the elimination of direct blockage.

Offset-fed dual-reflector antennas exhibit sidelobe performance similar to that of the prime-focus-fed offset reflector. Two offset-fed dual-reflector geometries have been used for earth station antennas: the double-offset geometry and the open Cassegrain geometry introduced by Cook, et al., of Bell Laboratories. In the double-offset geometry, the feed is located below the main reflector, and no blocking of the optical path occurs. By contrast, the

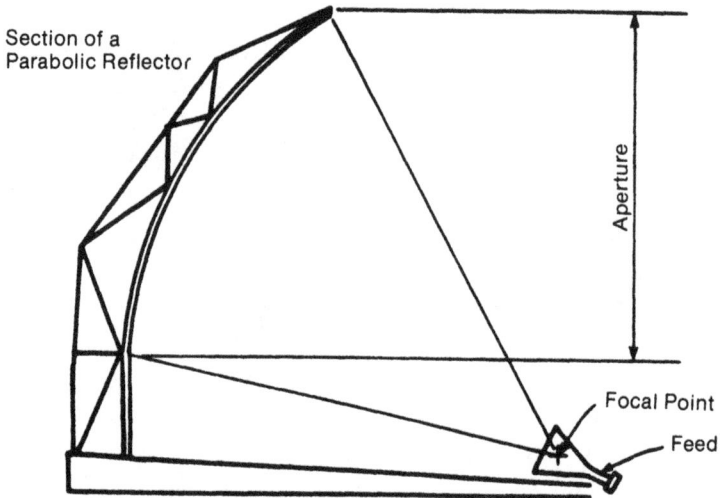

Figure 28.5 Basic Offset-Fed Parabolic Antenna
Source: Scientific Atlanta

open Cassegrain geometry is such that the primary feed protrudes through the main reflector; thus, it is not completely blockage free. Nevertheless, both geometries provide the capability of excellent sidelobe and efficiency performance.

The disadvantage of offset geometry antennas is that they are asymmetric. This leads to increased costs because there exists only one plane of symmetry. The geometry also has some effect on the electrical performance. The offset geometry, when used for linear polarization, has a depolarizing effect on the primary feed radiation and produces two cross-polarized lobes within the main beam in the plane of symmetry. When used with circular polarization, the geometry introduces a small amount of beam squint whose direction is dependent upon the sense of polarization (i.e., right- or left-handedness).

28.6 Multiple Beam Antennas

During the past few years there has been an increasing interest in receiving signals simultaneously from several satellites with a single antenna. This interest has prompted the development of several multibeam antennas, including the *spherical reflector* and *torus*.

28.7 Spherical Reflector

The properties, applications, and problems of the spherical reflector are familiar to microwave antenna designers. Its popularity is a result of the large angle through which the radiated beam can be scanned by translation and orientation of the primary feed. This wide-angle property results from the symmetry of the surface. Multiple beam operation is realized by placing multiple feeds along the focal surface. In the conventional use of the reflector surface, the minimum angular separation between adjacent beams is determined by the feed aperture size. The maximum number of beams is determined by the percentage of the total sphere which is covered by the reflector.

In practice, the individual feed of the spherical reflector illuminates a portion of the reflector surface such that a beam is formed coincident to the axis of the feed. (The conventional multibeam geometry is shown in Fig. 28.6.) All of the beams have similar radiation patterns and gains, although there is degradation in performance in comparison to the paraboloid. The advantage of this antenna is that the reflector areas illuminated by the individual feeds overlap, reducing the surface area for a given number of beams as compared with individual single beam antennas.

The scan angle of this geometry is dependent upon the reflector size. Figure 28.7 shows the conventional spherical reflector optics and the geometry for a single feed. The secondary beam is scanned as the feed is rotated around the feed scanning circle. A decrease in gain results, which is largely dependent on the f/D ratio.

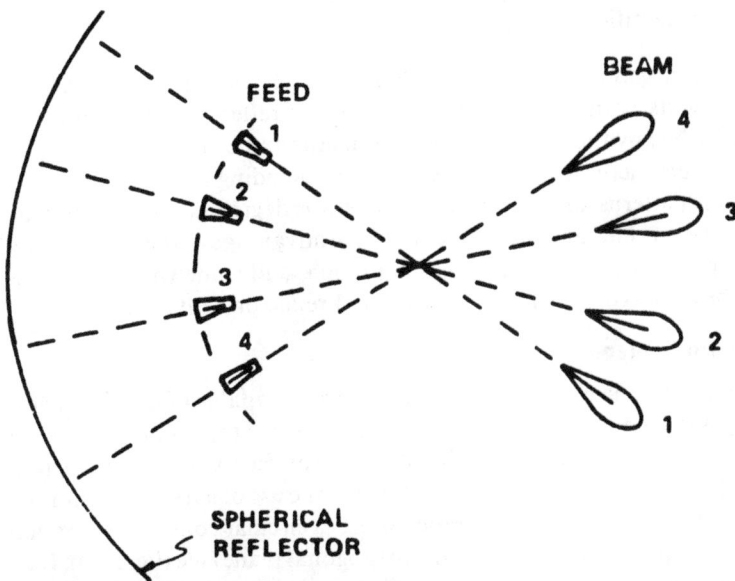

Figure 28.6 Conventional Spherical Multibeam Antenna Using Extended Reflector and Multiple Feed
Source: Scientific Atlanta

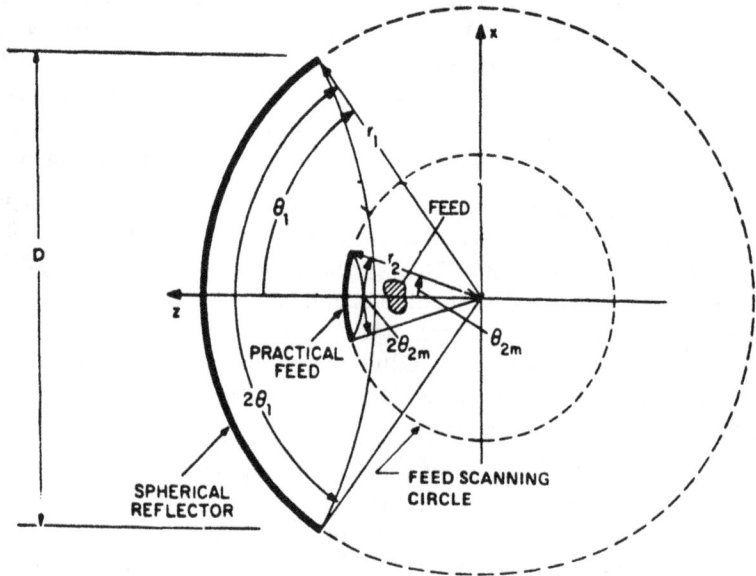

Figure 28.7 Convention Spherical Reflector and Feed Geometry
Source: Scientific Atlanta

An alternate geometry is shown in Fig. 28.8. For this geometry, each of the feed elements points toward the center of the reflector with beam steering accomplished by the feed position. This method of beam generation leads to considerable increase in *phase aberration*, including *coma*. As a result, the radiation patterns of the off-axis beams are degraded with respect to the on-axis beam. This approach does not take advantage of the spherical reflector properties of the conventional approach, and somewhat similar results could be achieved by using a paraboloidal reflector with a large f/D.

28.8 Torus Antenna

The torus antenna is a dual-curvature reflector, capable of multibeam operation by way feeding it with multiple feeds similar to those of the conventional spherical reflector geometry. The plane of the feed scan can be inclined to coincide with the orbital arc plane, allowing the use of a fixed reflector to view geosynchronous satellites. The reflector has a circular contour in the scanning plane and a parabolic contour in the orthogonal plane (see Fig. 28.9). It can be fed in either an axisymmetric or an offset configuration. The offset geometry for use as an earth station antenna has been successfully demonstrated by COMSAT laboratories. The radiation patterns meet the $(32-25 \log \theta)$ dBi envelope.

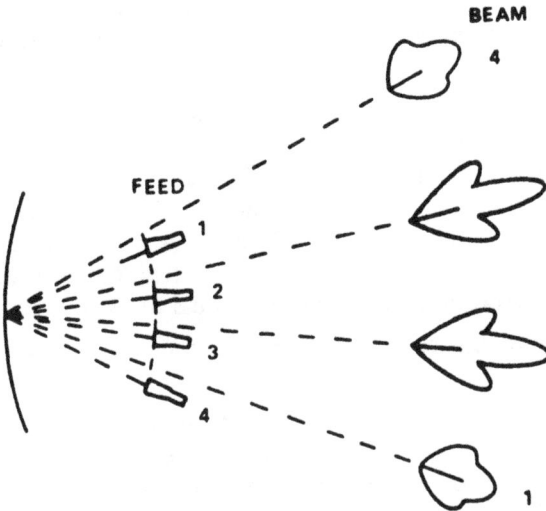

Figure 28.8 Alternate Spherical Multibeam Antenna Using Minimum Reflector Aperture with Scanned Beam Feeds
Source: Scientific Atlanta

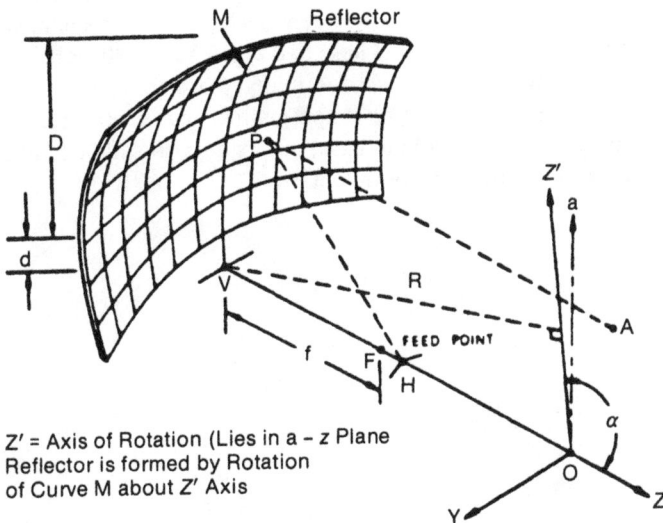

Z′ = Axis of Rotation (Lies in a – z Plane
Reflector is formed by Rotation
of Curve M about Z′ Axis

Figure 28.9 Torus Antenna Geometry
Source: Scientific Atlanta

The offset-fed geometry results in an unblocked aperture, which gives rise to low wide-angle sidelobes as well as providing convenient access to the multiple feeds.

The torus antenna has less phase aberration than the spherical reflector because of the focusing in the parabolic plane. By virtue of its circular symmetry, feeds may be placed anywhere along the feed arc from identical beams. As a result, no performance degradation is incurred when multiple beams are placed on the focal arc. Point-focus feeds may be used to feed the torus up to aperture diameters of approximately 200 wavelengths. For larger apertures, aberration-correcting feeds should be used.

Torus requires an oversized aperture to accomodate the scanning, or multi-beam, operation.

The main disadvantage of the torus antenna is in its stringent installation requirements. The mounting structure must be customized for the particular site latitude and longitude, because the reflector scanning plane must be accurately located relative to the orbital plane. A secondary structure or building is also required to support the multiple feeds and associated electronics.

Many kinds of antenna are used as earth station antennas. For overall performance, the prime-focus-fed paraboloidal and the dual-reflector Cassegrain antennas have been the predominant types used. Their choice has been based on the best compromise between electrical and mechanical performance commensurate with cost. These two types will likely continue to be the earth station antennas of choice for many years to come, although some increase in the use of offset-reflector geometries is expected. The use of multiple-beam antennas will increase, but its growth will be tempered by the added cost and installation complexity.

29   Data Communication

Data communications is a dynamic, rapidly expanding field, and cable television facilities are quite suitable for wideband data communications.* System operators can substantially increase their revenues by adding relatively low cost data communications capabilities to their existing facilities.

Digital computers and digital communications are based upon the principles of *binary logic*.** The presence or absence of a signal during a specific bit of time can be used to produce information. This is referred to as binary because only one of two conditions can exist during the specific time interval: signal or no signal; on or off; true or false; or, in data communications terminology, zero or one. All of these expressions represent binary states.

In data communications language the zero or one is referred to as a *bit*. Bit is a contraction of the words *binary*, meaning something having two parts or two possible states, and *digit*, meaning number. A bit is a zero or a one, and is the smallest data unit. Binary arithmetic uses a system of numbers having two as a base.

Switches, relays, transistors, flip-flops, and similar electronic components are all binary devices. If information is converted into binary digits, these components can be used to communicate.

If more than one bit is used, the first bit gives two possible conditions, zero or one. Adding a second bit gives two more for a total of four possibilities. A third bit increases the possibilities to eight.

---

*See Ralph Glasgal, *Techniques in Data Communications*, (Dedham, MA: Artech House, 1983).

**Aristotle developed a logic system to explain his philosophical concepts. This system used statements which were either true or false. In 1847, George Boole developed Boolean algebra, which reduces the Aristotlean system to a universal logic language called binary logic.

Bit Number	1	2	3	4	5	6	7	8
Value	2^1	2^2	2^3	2^4	2^5	2^6	2^7	2^8
	2	4	8	16	32	64	128	256

Four bits produce 16 possibilities and five bits produce 32. Table 29.1 shows the progression of possibilities as another bit is added. Note that each successive bit represents a higher power of two ($2^0 = 1$; $2^1 = 2$; $2^2 = 4$; $2^3 = 8$; *et cetera*. This is the same progression as in binary arithmetic.

Because five bits give a total of 32 possibilities, a five-bit code can easily be used to represent the 26 characters of the alphabet, with each character having its own combination of zeros and ones. One type of teletype operation is designed around a five-bit binary code known as *Baudot* (named after Jean M.E. Baudot, a Frenchman who made many contributions to early telegraph principles).

Figure 29.1 shows the Baudot code for several characters and the pulse widths for different commonly used telegraph speeds. When current is flowing in the teletype circuit (the shaded portions of Fig. 29.1), it is called a *mark*. A bit in which no current is flowing (unshaded) is called a *space*.

As shown in Fig. 29.1, there is a *start pulse* at the beginning of each character. In start/stop teletype code, the pulse is always a space. The next five pulses can be either marks or spaces depending on what character is being sent. The last pulse is a *stop pulse* which is always a mark condition.

If the timing of the bits is tightly controlled, the start and stop pulses for each character can be eliminated. This type of transmission is known as *synchronous*. By eliminating the start and stop pulses, synchronous transmission enables more characters to be sent in the same amount of time, or a greater number of bits to be used to represent a character. The timing circuit used to control synchronous transmission is called a *clock*, and is very important to computers and most digital transmission systems.

29.1 Parity Bits

If a sixth bit is added to the five-bit character, the code can be used to represent 64 different possibilities, or the sixth bit can be used to check for errors in transmission. This is called a *parity check*, and the sixth bit is the *parity bit*. The parity bit can be used to check for errors in many ways. For example, Fig. 29.1 shows that the total mark bits in any character can be an odd or even number. Suppose a system rule is established that every character

LET-TER	START	1	2	3	4	5	STOP
A		▓					▓
H				▓		▓	▓
R			▓		▓		▓
Y		▓		▓		▓	▓

Code Pulse Positions

60 WPM	22	22	22	22	22	22	31
75 WPM	18	18	18	18	18	18	25
100 WPM	13.5	13.5	13.5	13.5	13.5	13.5	19

Pulse Duration Milliseconds

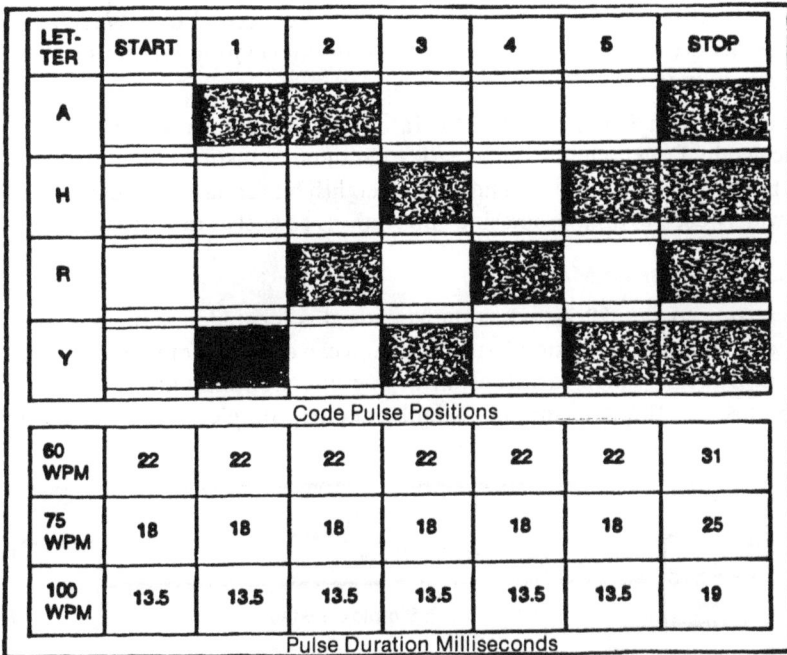

Figure 29.1 Example of 5-Level Baudon. Note: code mark pulses are shaded, space pulses are blank.

must have an even number of mark bits. The letter Y has three mark bits. However, the rule states that every character must have an even number of marks, so the parity bit is made a mark to give four. The letter A has two marks, an even number, so the parity bit is made a space. Then, if a character with an odd number of marks is received, it must be in error.

29.2 Bits and Bauds

Bits and *bauds* are two data transmission terms which sometimes cause confusion. Bits per second (b/s) expresses the total number of information pulses in one second and includes redundant bits, such as parity bits, for checking errors. The baud, named after the same Monsieur Baudot mentioned earlier, is defined as the reciprocal of the shortest signal element in a character.

Referring to the 100 word-per-minute (wpm) speed in Fig. 29.1, the shortest signal element in the five-bit character is 13.5 ms, or approximately 74.1 bauds. A single character consists of a start pulse and five information pulses, each 13.5 ms in duration, for a total of 81 ms. A stop pulse of 19 ms ends the character, so the total time for a character is 100 ms.

Because baud speed depends on the number of information pulses, the number of bauds equals the number of bits per second only when all time intervals are constant and every pulse is an information bit. This condition occurs in many forms of binary codes.

The American Standard Code for Information Interchange (ASCII) is one such code. Developed by the American Standards Institute, it uses seven bits to provide 128 possibilities and adds an eighth bit for parity checking. ASCII is widely used in data processing applications as well as in data transmission.

29.3 Transmission Modes

Referring to Fig. 29.2a, when the keyboard is located at station A and the printer is located at station B, transmission can only occur in one direction, A to B. Systems which can transmit in one direction only are called *simplex* systems; i.e., they transmit in a simplex mode. Simplex systems are seldom

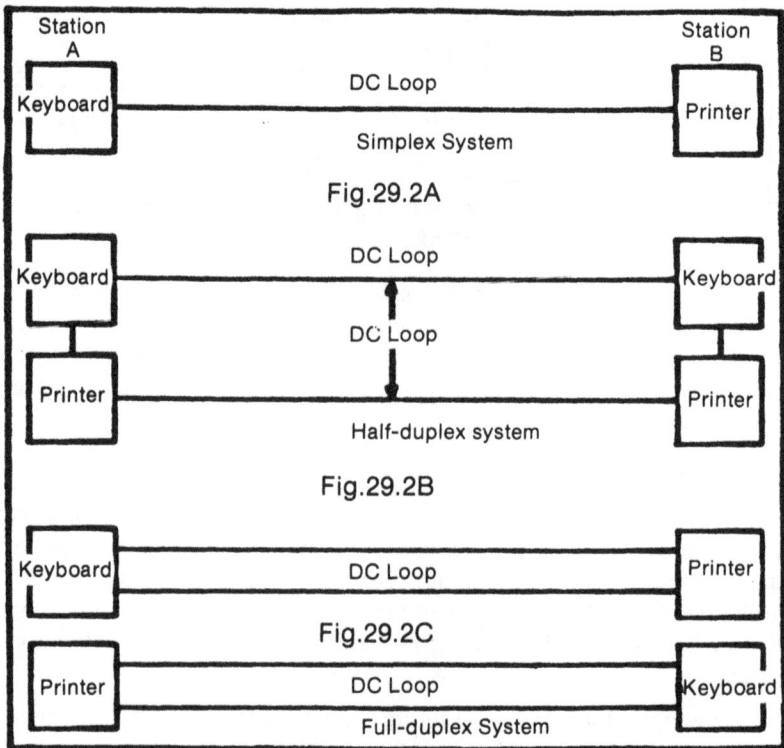

Figure 29.2 (a) Simplex system. (b) Half-Duplex system. (c) Full-Duplex system.

used for data communications because it is not possible to return error or control signals from the receiving end to the transmitting end.

When a keyboard and printer are available at both ends, and interconnected in the same dc loop as shown in Fig. 29.2b, both ends can transmit, but not simultaneously. This is known as a *half-duplex* system, which is probably the most common transmission mode in use today.

When the keyboard is at station A in a dc loop with a printer at station B, and the keyboard at station B is in a separate loop with the printer at station A — as shown in Fig. 29.2c — transmission in both directions can occur simultaneously. This is known as *full-duplex* operation.

The full-duplex mode provides the most efficient data transmission or "through-put" rate. In data communications, efficiency ("through-put" rate) is defined as the number of bits received divided by the number of bits transmitted. *Data communications equipment* (DCE) and *data terminal equipment* (DTE) are described by their ability to operate in one or more of the three transmission modes.

29.4 Modulation

Characters and symbols can be converted into bits and transmitted as current pulses over a dc circuit; but, for transmitting data over telephone voice, radio, or CATV circuits, the data must be changed to a form that the transmission medium can pass.

A *modem* (contraction of the words *modulator* and *demodulator*) is one device used to accomplish this change. A modem accepts digital data and translates it to an analog signal suitable for transmission over a particular type of network. A modem also receives analog signals from other modems and translates them to a digital from suitable for input to a computer, data terminal, printer, *et cetera*.

Three techniques are used to convert digital data into analog signals: *amplitude shift keying* (ASK), *frequency shift keying* (FSK), and *phase shift keying* (PSK). In ASK, data signals control the amplitude of the sine wave carrier; in FSK, data signals control the frequency of the carrier; and PSK varies the phase angle. Figure 29.3 illustrates the three conversion methods.

When data is transmitted at relatively low-speed rates, the modulated carrier can carry information in a binary form. For example, an ASK signal may be at a high or low level, an FSK signal may be one of two distinct frequencies, and a PSK signal can be either unchanged or shifted 180°. The information content of these signals is one bit per pulse.

Expressed in bits, the information content is $H - \log_2 m$,

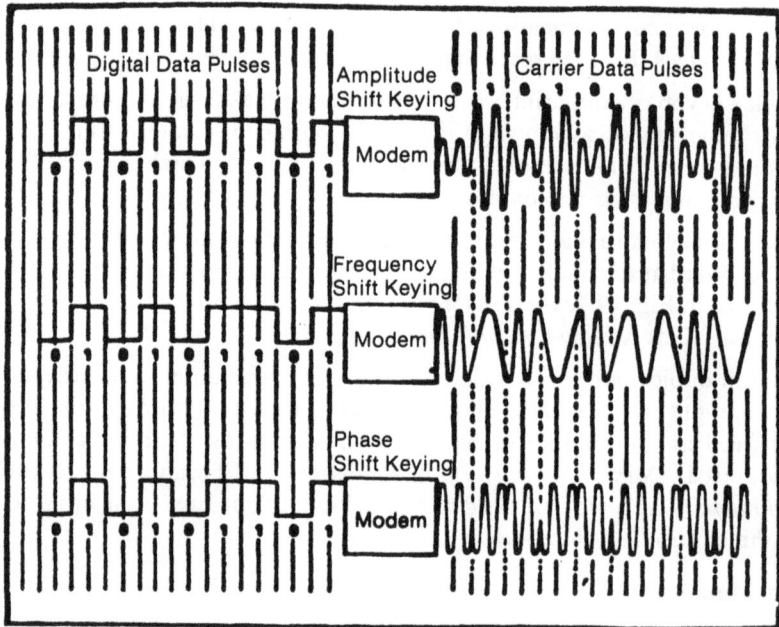

Figure 29.3 Methods for Converting Data Pulses to a Form Suitable for Analog Transmission.

where

H = number of bits
m = number of choices

Because there are two choices ($m = 2$) in the binary format and $\log_2 2 = 1$, one bit of information can be conveyed.

29.5 Bandwidth

Many studies have shown that a transmission medium with an *ideal bandpass* can accommodate a maximum number of carrier pulses equal to the bandwidth in hertz. An ideal bandpass is rectangular in shape and all frequencies within the band are passed with the same minimum attenuation. However, an ideal bandpass cannot be physically realized, and transmission paths contain noise and distortion, hence the number of transmission pulses a path can accommodate may thus be significantly less than the bandwidth in hertz. Therefore, non-binary coding techniques are used for systems transmitting high-speed data to make each pulse contain more than a single bit rate.

Voice-band modems range in speed from as low as 300 bps up to 19.2 kilobits per second (kbps). One kilobit is 1000 bits. Voice-band modems are the most commonly used to access the telephone network. The term *voice-band* relates to this, not to transmission speed. Voice-band modems are subdivided into low speed (below 1200 bps), medium speed (1200 to 2400 bps), and high speed (up to 19.2 kilobits).

Wideband modems are high-speed devices. They are also sub-divided into three categories: the first category includes speeds from around 20 kpbs to 50 kbps; the second goes up to about 500 kbps; and the third, the video category, goes up to 20 megabits per second (mbps). A megabit is one million bits.

29.6 Multiplexing

Data multiplexing is a way of using a common path to transmit a number of data channels. Two methods are used: *frequency-division multiplexing* (FDM) and *time-division multiplexing* (TDM).

FDM divides the total transmission frequency bandwidth into narrower data bands, each used as a separate data channel. All of the users share the facility simultaneously, with a portion of the total allocated transmission spectrum always available to each user.

TDM enables the transmission of more than one data channel over a single path by using a switching arrangement to transmit each channel sequentially at different instants. The channel is available to only one user at a time. Figure 29.4 shows two methods.

29.7 Series and Parallel

Series (or serial) and *parallel* are two techniques used to transmit data. Information transmitted serially is handled sequentially, bit by bit, in a single file. Information transmitted in parallel is formed into characters, words, or blocks which are transmitted simultaneously.

A common type of business machine generates eight bits of data and presents them simultaneously on eight separate outputs. Each parallel set of eight bits represents a character, so the output is called *parallel by bit, serial by character*.

Business machines with parallel outputs can use parallel or series data trans-mission. For serial transmission, a *parallel-to-series converter* is used to interface between the business machine and the series data transmission. At the receiving terminal, another converter is used to convert the serial data · back to a parallel format.

Both serial and parallel transmission systems have inherent disadvantages. The serial system requires additional converters for parallel interface. On the other hand, the parallel transmitter must provide several oscillators for multiplexing the side-by-side channels, making parallel transmission more susceptible to frequency error.

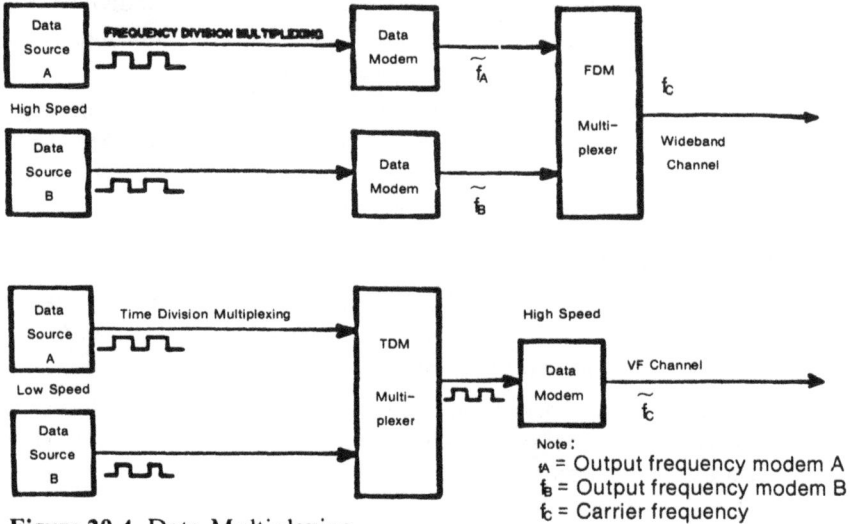

Figure 29.4 Data Multiplexing

Note:
f_A = Output frequency modem A
f_B = Output frequency modem B
f_C = Carrier frequency

Furthermore, parallel transmission requires that a portion of the available bandwidth be used to provide guard bands between the parallel channels, but serial systems can use the entire linear portion of the available bandwidth to transmit data. The choice between series and parallel transmission depends on the transmission speed required and the data processing equipment to be served.

29.8 Compatibility

It is essential that the data terminal equipment and data communications equipment be compatible. Several standards have been established to ensure this compatability. Organizations interested in establishing these standards include the International Telephone and Telegraph Consultative Committee (CCITT), the Institute of Electrical and Electronic Engineers (IEEE), and the Electronics Industry Association (EIA).

Most North American data communications equipment, using a voltage interface, is designed to conform to Electronics Industry Association standard RS-232C. (In general usage, the interface is called RS-232, dropping the letter C at the end.) The standard defines a binary one as any interface voltage between –3 and –25 Vdc. A binary zero is any voltage between +3 and +25 Vdc. The standard also specifies the signal and control functions assigned to each of the 25 pins in an RS-232C connector.

As the field of cable television technology continues to grow, data communications will play a large role in that growth.

Appendix

List of Symbols Used in CATV

——✕——	Power Pole
——◯——	Telephone Pole
——⊗——	Power and Telephone, Joint Use
——⊠——	Joint Use Pole with Transformer
——⊘——	CATV Pole
——●—— ⋯◌⋯	Proposed Pole
——⊗₆₃₂₁₃₄—	Pole Number
⊗————⊗	Tensioned Messenger Wire
⊗⌣⊗	Slack Span
⊗—⊣⊢—⊗	Overhead Guy
⊗—▶—⊗	Pole to Pole Guy
⊗⌐↓	Sidewalk Down Guy
⊗⌐✗	Sidewalk Down Guy with Anchor
⊗—▶	Down Guy
⊗—▶◄	Down Guy with Anchor

Trunk Amplifier with ALC

Trunk Amplifier with Bridger

Terminating Bridger

Line Extender

AC Power Supply

Standby Power Supply

Two-way Splitter

Three-way Splitter, dot on High Leg

Directional Coupler 8 dB

Directional Coupler 12 dB

Directional Coupler 16 dB

AC Block

AC Block and Two-Way Splitter

Fixed Equalizer

Variable Equalizer

Line Terminator

Two-Output Tap

Four-output Tap

Eight-Output Tap

Apartment Building with Number of Units

Trunk Amplifier with Bridger

Trunk Line

Feeder Line

Line Extender with 8 dB Directional Coupler

Line Extender with Two-way Splitter

Glossary

Active Video Lines

    All video lines not occurring in the vertical blanking interval

A/D

    Analog-to-digital converter

AFC

    Automatic frequency control

AGC

    See automatic gain control

AISC

    American Institute of Steel Construction

ALC

    Automatic level control. *See* automatic gain control

ANSI

    American National Standards Institute

Antenna Preamplifier

    A small amplifier located in the immediate vicinity of the antenna, used to amplify extremely weak signals, thereby improving the signal-to-noise ratio of a system

ASCII

American Standard Code for Information Interchange

ASMS

Automatic surface measuring system

ASTM

American Society for Testing and Materials

Automatic Gain Control (AGC)

A circuit which automatically controls the gain of an amplifier so that the output signal level is virtually constant for varying input signal levels. Sometimes referred to as automatic level control (ALC) or automatic volume control

Automatic Temperature Control

A method whereby changes due to temperature in amplifiers or coaxial cable are automatically corrected by either a closed or open servo system

Automatic Tilt

Automatic correction of changes in tilt

Average Picture Level (APL)

The average luminance level of the unblanked portion of a television line measured in IRE standard units

AZ

Azimuth angle

Back Porch

That portion of the composite video signal which lies between the trailing edge of the horizontal synchronizing pulse and the trailing edge of the horizontal blanking pulse

BER

Bit error rate

Bit

Binary digit

Black Level

The instantaneous amplitude of the television signal which corresponds to a black area in the received picture

Black Peak

The maximum excursion of the picture signal in the black direction at the time of observation

Blanking

The process of cutting off the electron beam in a camera or picture tube during the retrace period

Blanking Level

The level of the front and back porches of the composite video signal

Breezeway

The portion of the back porch between the trailing edge of the synchronizing pulse and the start of the color burst

Bridging Amplifier (Bridger)

An amplifier which is connected directly into the main trunk of a CATV system, providing isolation between the main trunk and multiple high level outputs

Burst Flag

Pulses used to key out a portion of the 3.579545 MHz sine wave subcarrier for use as a reference for the color signal

Byte

Grouping of bits, usually eight

Cable Powering

A method of supplying power by utilizing the coaxial cable to carry both signal and power simultaneously

CATV

Community Antenna Television, synonymous with cable TV

CCIR

International Radio Consultative Committee

CCTV

Closed circuit television

Chrominance

That property of light which produces a sensation of color in the human eye apart from any variation in luminance that may be present

Clamper

A device which functions during the horizontal blanking or synchronizing interval to fix the picture signal at some predetermined reference level at the beginning of each scanning line

Clamping

The process that establishes a fixed picture level at the beginning of each scanning line

Clipping

The shearing off of the peaks of a signal. For a picture signal, this may affect either the positive (white) or negative (black) peaks. For a composite video signal, the synchronizing signal may be affected

Closed Loop System

A servo feedback system where the residual error after correction is fed back directly into the servo system for inverse proportional control

C/N

Carrier-to-noise ratio

C/N_o

Carrier-to-noise power density ratio

Coaxial Cable

The most commonly used means of signal distribution, consisting of a center conductor and a cylindrical outer conductor (shield).

Color Bar

A test signal, typically containing six basic colors: yellow, cyan, green, magenta, red, and blue; used to check the chrominance functions of color TV systems

Color Burst

This normally refers to a burst of approximately 10 cycles of 3.579545 MHz subcarrier frequency of the back porch of the composite video signal. This serves as a color synchronizing signal to establish frequency and phase references for the chrominance signal

Color Contamination

An error of color rendition due to incomplete separation of paths carrying different color components of the picture

Color Decoder

In color television reception, an apparatus for deriving the receiver primary signals from the color picture signal and the color burst

Color Phase

The phase, with respect to the chrominance carrier reference, of that component of the chrominance carrier signal which corresponds to one of the chrominance primaries

Color Signal

Any signal, excepts for the luminance of monochrome signal, at any point in a color television system, for wholly or partially controlling the chromaticity of a color television picture. This is a general term which encompasses many specific connotations

Color Subcarrier

In color systems, this is the carrier signal whose modulation sidebands are added to the monochrome signals to convey color information; a 3.579545 MHz sine wave

Combining Network

A passive network which permits the addition of several signals into one combined output with a high degree of isolation between individual inputs

Compatibility

That property of a color television system which permits substantially normal receivers

Composite Blanking

A signal composed of pulses at line and field frequencies used to make the return traces of a picture tube invisible

Composite Synchronization

The line and field rate synchronizing pulses (including the field equalizing pulses) when combined together

Composite Video

For color, this consists of blanking, field, line, and color synchronizing signals; chrominance and luminance picture information; combined to form the complete color video signal

Compression

The reduction in gain at one level of a picture signal with respect to the gain at another level of the same signal

Contrast

The ratio between the maximum and minimum brightness values in a picture

CONUS

Contiguous (48) United States, or continental United States

Convergence

In color television, the meeting or crossing of the three electron beams at the shadow mask

Converter

A device to convert one or more television channels to one or more other channels

Cross-Hatch

A grid of vertical and horizontal white bars over a black background

Cross Modulation

A form of distortion where modulation of an interfering station appears as a modulation of the desired station, caused by third-order nonlinearities

D/A

Digital-to-analog converter

DC Restorer

A device, used in an ac transmission system to establish dc transmission

DEC

Declination angle

Decibel (dB)

A term that expresses the ratio of two power levels used to indicate gains or losses in a system. Also used to express absolute power levels such as dBm, dBW, dBmV

De-emphasis

The restoration of a pre-emphasized signal to its original wave form (*see* pre-emphasis)

Delay Distortion

That form of distortion which occurs when the envelope delay of a circuit or system is not constant over the frequency range required for transmission

Differential Gain

The amplitude change, usually of the 3.579545 MHz color subcarrier, introduced by the overall circuit measured in degrees as the picture signal on which it rides is varied from blanking to white level

Differential Coupler

A device having one input and providing two or more isolated outlets for RF cable runs; also called line splitter.

Distribution System

The part of a CATV system used to carry signals from the head-end to subscribers' receivers. Often applied, more narrowly, to the part of a CATV system starting at the bridger amplifiers

DMUX

Demultiplex (*see* multiplex)

Downlink

The circuit between a satellite and a receiving earth station

E_b/N_o

Energy per Bit to Noise Power Density Ratio

Echo

A signal which has been reflected at one or more points during transmission with sufficient magnitude and time difference to be detected as a signal distinct from the primary signal. Echoes can be either leading or lagging compared to the primary signal and appear as reflection or "ghosts"

EIA

Electronic Industries Association

EIRP

Effective isotropic radiated power product of transmitted power times transmitting antenna gain, usually expressed in dBW

EL

Elevation angle

Envelope Delay

The first derivative of the phase shift with reference to the frequency

ET

Earth terminal

Equalizing Pulses

Pulses of one-half the width of the horizontal synchronizing pulses which are transmitted at twice the rate of the horizontal synchronizing pulses during the portions of the vertical blanking interval immediately preceding and following the vertical synchronizing pulses. The purpose of these pulses is to cause the vertical deflection to start at the same time in each interval, and to keep the horizontal sweep circuits in step during the portions of the vertical blanking interval immediately preceding and following the vertical synchronous pulse

Expansion

An undesired increase in the amplitude of a portion of the composite video signal relative to another portion, or a greater than proportional change in the output of a circuit for a change in input level. Opposite of compression

Fader

A control or group of controls for fade in and fade out of video or audio signals

FEC

Forward error correction. An encoding scheme to improve the BER in a data system. In a typical rate ¾ FEC system, the BER may be improved by a factor of ten thousand

Feeder Line

The coaxial cable running between bridgers, line extenders, and taps

Field

One-half of a complete picture (or frame) interval, containing all of the odd, or all of the even, lines of the picture

Field Blanking

Refers to the blanking signals which occur at the end of each field; also called vertical blanking

Field Frequency

The rate at which one complete field is scanned, normally 59.54 times per second

Frame

One complete picture consisting of two fields of interlaced scanning lines

Frequency-Division Multiplex (FDM)

A method of multiplexing or combining many voice data channels for transmission on a single RF carrier. The channels are separated in frequency and are carried on subcarriers

Frequency-Division Multiple Access (FDMA)

Frequency division capacity of a satellite transponder allowing access by multiple earth stations

Frequency Modulation (FM)

A system of modulation where the instantaneous radio frequency varies in proportion to the instantaneous amplitude of the modulating signal (amplitude of modulating signal to be measured after pre-emphasis), and the instantaneous radio frequency is independent of the frequency of the modulating signal. Angle modulation in which the instantaneous frequency of a sine wave carrier is caused to depart from the carrier frequency by an amount proportional to the instantaneous value of the modulating wave

Frequency Response

The change of gain with frequency

Frequency Re-use

A technique in which independent information is transmitted on orthogonal polarizations to re-use a given band of frequencies

Front Porch

That portion of the composite picture signal which lies between the leading edge of the horizontal blanking pulse and the leading edge of the corresponding synchronizing pulse

Gain

A measure of amplification usually expressed in dB. For matched CATV components, power gain is readily determined as insertion power gain. Gain of an amplifier is often specified at the highest frequency of operation

GCE

Ground communications equipment. This term relates to the earth station electronic equipment, such as receivers and exciters

Ghosts

Ghosts are pictures, either positive or negative, displaced in time from the desired picture due to multipath transmission or reflections in the apparatus. A ghost displaced to the left of the primary image is designated as leading and one displaced to the right is designated as following (lagging). When the tonal variations of the ghost are the same as those of the primary image it is designated as positive and when the reverse condition occurs it is designated as negative. (*See* echo)

G/T

Figure of merit of a receiving system, expressed in dB/K. G is the net gain of the antenna referenced to the point of measurement and T is the noise temperature of the system in degrees Kelvin referenced to the same point. The value of G/T is independent of the point at which it is measured

HA

Hour angle

Head-end

The electronic equipment located at the start of a cable system, usually including antennas, preamplifiers, frequency converters, demodulators, modulators, and related equipment

High-band

The higher VHF television channels, 7 through 13

Horizontal Blanking

The blanking signal at the end of each scanning line

Horizontal Drive

A pulse at H-rate in TV cameras. Its leading edge is coincident with the leading edge of the horizontal synchronizing pulse and the trailing edge is coincident with the leading edge of the burst flag pulse

Horizontal (Hum) Bars

Relatively broad horizontal bars, alternately black and white, which extend over the entire picture. They may be stationary or may move up and down. Sometimes referred to as a "venetian blind" effect, it is usually caused by a 60 Hz interfering frequency or a harmonic frequency

House Drop

The coaxial cable from line tap to subscriber's TV set

HPA

High-power amplifier. In a transmitting earth station, this is the final RF amplifier between the modulator/exciter and the antenna

H Rate

The time for scanning one complete line, including trace and retrace; equals 1/15734 second (color) or 63.56 microseconds

Hue

The attribute of color perception that determines whether the color is red, yellow, green, blue, *et cetera*. White, black, and gray are not hues

In-line Package

A housing for amplifiers or other CATV components designed for use without jumper cables; cable connectors on the housing ends are in line with the coaxial cable

Insertion Loss

Additional loss in a system when a device such as a directional coupler is inserted; equal to the difference in signal level between input and output of such a device

Intercity Relay System

A system for the transmission of television relay signals by fixed stations within a given city or service area for purposes other than those of a studio-to-transmitter relay system

IRE

Institute of Radio Engineers

IRE Scale

An oscilloscope scale that applies to composite video levels. There are 140 IRE standard units in 1 volt

Kelvin (K)

Temperature of a device measured in Kelvin degrees. Zero K equals —273°C or —459°F. The Kelvin scale is the same as the Celsius, or centigrade, scale except for the offset of 270°K

Level Diagram

A graphic diagram indicating the signal level at any point in a system

LHA

Local hour angle

Line Amplifier

An amplifier installed at an intermediate position connected to a main cable run in a master antenna or CATV system to compensate for loss; generally a broadband amplifier

Line Blanking

The blanking signal at the end of each scanning line used to make the horizontal retrace invisible; also called horizontal blanking

Line Extender

Type of amplifier used in the feeder system

Line Frequency

The number of horizontal scans per second, normally 15,734.26 times per second

LNA

Low-noise amplifier. This is the preamplifier between the antenna and the earth station receiver. For maximum effectiveness, it must be located as near the antenna as possible and is usually attached directly to the antenna receiving port

Low-Band

The lower VHF television channels, 2 through 6

Low-Frequency Distortion

Distortion effects which occur at low frequency. In television, generally considered as any frequency below the 15.75 kHz line frequency

Luminance

The amount of light intensity which is perceived by the eye as brightness, referred to as Y

Luminance Channel

In a color television system any path which is intended to carry the luminance signal

Main Trunk

The major link from the head-end to a community or connecting communities

Master Antenna Television System (MATV)

A combination of components providing multiple television receiver operation from one antenna or group of antennas; normally within a single building

Modem

Modulator-demodulator. Usually a device that combines the modulation and demodulation functions in a single unit

MTBF

Mean time between failure. A statistical determination of the time, in hours of use, between failures

MUX

Multiplexer

Negative Image

A picture signal having a polarity which is opposite to normal polarity, resulting in a picture in which the white and black areas are reversed

Negative Modulation

In an AM television system, that form of modulation in which an increase in brightness corresponds to a decrease in transmitted power. Opposite of positive modulation

Noise Figure (NF)

A figure of merit of a device, such as an LNA or receiver, which compares the device with a perfect device. The noise figure or noise factor of a device, expressed in dB, is related to its noise temperature

Noise Power (NP)

A measure of the noisiness of an amplifier. Noise power is defined as input signal-to-noise ratio to output signal-to-noise ratio. Noise power is expressed in dB and the lowest possible value for a matched system is 3 dB

NTSC

National Television Systems Committee. An industry-wide engineering group which, during 1950-1953, developed the color television specifications now established in the U S

OMT

Orthomode transducer. A device attached to an antenna feed that permits using the antenna for simultaneous transmission and reception. The device is polarization selective; the transmit and receive signals must be on orthogonal polarizations

Overshoot

The initial transient response to a unidirectional change in input, which exceeds the steady-state response

Pairing

The overlapping of alternate scanning lines resulting in a severe reduction in vertical resolution capability

Polarization

The direction of the electric vector of the radiated signal

Positive Modulation

In an AM television system, that form of modulation in which an increase in brightness corresponds to an increase in transmitted power. Opposite of negative modulation

Pre-emphasis

The intentional alteration of the normal signal wave by emphasizing one range of frequencies with respect to another

PSK

Phase shift keying

QPSK

Quadriphase shift keying

Random Interface

A technique for scanning often used in closed-circuit television systems, providing somewhat reduced precision compared to that used in commercial broadcast service

Raster

A predetermined pattern of scanning lines which provides substantially uniform coverage of an area

RCVR

Receiver

Receiver Primaries

The colors of constant chrominance and variable luminance produced by the receiver which, when mixed in proper proportions, are used to produce other colors. Usually three primaries are used: red, green, and blue

Reference Black Level

The level corresponding to the specified maximum excursion of the luminance signal in the black direction. The level at the point of observation corresponding to the specified maximum excursion of the picture signal in the black direction

Reference White Level

The level at the point of observation corresponding to the specified maximum excursion of the picture signal in the white direction

Resolution (horizontal)

The amount of resolvable detail in the horizontal direction in a picture. It is usually expressed as the number of distinct vertical lines, alternately black and white, which can be seen in a distance equal to picture height

Resolution (vertical)

The amount of resolvable detail in the vertical direction in a picture. It is usually expressed as the number of distinct horizontal lines, alternately black and white, which can theoretically be seen in a picture

Return Trace

The path of the scanning spot during the return interval

Ringing

An oscillatory transient occuring in the output of a system as a result of a sudden change in input

RMS

Root mean square

Roll

A loss of vertical synchronization which causes the picture to move up or down on receiver or monitor

Saturation

This indicates how little a color is diluted by white light, distinguishing between vivid and weak shades of the same hue. The more a color differs from white, the greater is its saturation. Saturation is also indicated by the terms purity and chroma. High purity and chroma correspond to high saturation and vivid color

SCPC

Single channel per carrier. A satellite transmission system that employs a separate carrier for each channel. This method is usually used at earth stations that have a low volume of traffic compared to that which can be carried by a full transponder

Setup

The separation in level between blanking and reference black levels

Signal Strength

The intensity of the television signal measured in volts, millivolts, micro-volts, or dB, using 0 dB as a reference; equal to 1000 microvolts in RF systems; generally 1 volt in video systems

S/N

Signal-to-noise ratio. The ratio of signal power and noise power. A video S/N of 54 to 56 dB is considered to be excellent; that is, of best broadcast quality. A video S/N of 48 to 52 dB is considered to be good at the head-end for cable TV

Snow

Heavy random noise

Solid State

A term taken from physics, used interchangeably with the word transistor-ized; also includes other semiconductor elements, such as diodes. General-ly refers to tubeless equipment

Spacing

Length of cable between amplifiers based on the amount of gain required to overcome cable losses in dB at the highest TV channel carried in a system; usually at Channel 13 in an all-band system; equal to the dB loss between amplifiers at the same frequency

Splitter

A network supplying a signal to a number of outputs which are individually matched and isolated from each other; usually based on hybrid coils

Staircase

A video test signal containing several steps at increasing luminance levels. The staircase signal is usually amplitude modulated by the subcarrier frequency and is useful for checking amplitude and phase linearities in video systems

Streaking

A term used to describe a picture condition in which objects appear to be extended horizontally beyond their normal boundaries

Sub-Low Channels

TV channels between 5 MHz and 54 MHz

Sync

An abbreviation for the words synchronization, synchronizing, synchronous. Applies to the synchronization signals, or timing pulses, which lock the electron beam of the picture monitors in step, both horizontally and vertically, with the electron beam of the pickup tube. The color sync signal is known as the color burst

Sync Compression

The reduction in the amplitude of the sync signal, with the respect to the picture signal, occurring between two points of a circuit

Sync Level

The level of the peaks of the synchronizing signal

Synchronizing Signal

That portion of a composite picture signal which is more negative than the voltage representing the blanking level across the output of the picture line amplifier

Tap

Any device used to obtain signal voltages from a coaxial cable. The earlier forms such as capacitive and transformer taps have been replaced by directional couplers in modern system

TDM

Time-division multiplexing. (*See* also FDM)

TDMA

Time-division multiple access (of a satellite transponder by different earth stations)

Tearing

A term used to describe a picture condition in which groups of horizontal lines are displaced in an irregular manner

TED

Threshold extension demodulator. A scheme for lowering the threshold of FM demodulators, permitting operation of the system with a lower C/N without impulse noise showing on the picture

Television Channel

A band of frequencies 6 megacycles wide in the television broadcast band and designated either by number or by the extreme lower and upper frequencies

Termination

A term used in reference to impedance matching the ends of cable runs by installing a non-inductive resistor having the same resistance value as the characteristic impedance value of the cable

Test Pattern

A chart prepared for checking the overall performance of a television system, containing various combinations of lines and geometric shapes. The camera is focused on the chart and the pattern is viewed at a monitor

Tilt

The slope or change in cable attenuation or amplifier gain between Channels 2 and 13

Tilt-Compensation

The action of a compensated gain control whereby amplifier response is adjusted to provide the correct cable equalization for different lengths of cable; normally specified by range and tolerance

Transients

Signals which exist for a brief period of time prior to the attainment of a steady-state condition. These may include overshoots, damped sinusoidal waves, *et cetera*

Trap

A selective circuit used to attenuate undesired signals while not affecting desired signals

TVRO

Television receive-only (earth station)

TX

Transmit or transmitter

Ultra High Frequency (UHF)

In television, a term used to designate channels 14 through 83

Undershoot

The initial transient response to a unidirectional change in input which precedes the main transmission and is opposite in sense

Uplink

The circuit between a transmitting earth station and a satellite

Velocity of Propagation

Velocity of signal transmission. In free space, electromagnetic waves travel with the speed of light. In coaxial cables, this speed is reduced. Commonly expressed as percentage of the speed in free space

Vertical Blanking Interval

The blanking portion at the beginning of each field, containing the equalizing pulses, the vertical synchronizing pulses, and VITS

Vertical Drive

A pulse at field rate used in TV cameras. Its leading edge is coincident with the leading edge of the vertical blanking pulse and its duration is 10 ½ lines

Very High Frequency (VHF)

In television, a term used to designate channels 2 through 13

Vestigial Sideband

The transmitted portion of the sideband which has been largely suppressed by a transducer having a gradual cutoff in the neighborhood of the carrier frequency, the other sideband being transmitted without much suppression

Video

A term pertaining to the bandwidth and spectrum position of the signal resulting from television scanning. In current usage video means a bandwidth on the order of megacycles, and a spectrum position that goes with a dc carrier

Visual Carrier Frequency

The frequency of the carrier which is modulated by the picture information

VITS

Vertical interval test signal. A signal which may be included during the vertical blanking interval to permit on-the-air testing of video circuitry functions and adjustments

VSWR

Voltage standing wave ratio. Reflections present in a cable due to mismatch (faulty termination) combine with the original signal to produce voltage peaks and dips by addition and subtraction. The ratio of the peak-to-dip voltage is termed VSWR.

Wave Form Monitor

An oscilloscope designed especially for viewing the wave form of a video signal

White Clipper

A circuit designed to limit white peaks to a predetermined level

Windshield Wiper Effect

Onset of overload in multichannel CATV systems caused by cross modulation, where the horizontal synchronizing pulses of one or more TV channels are superimposed, out of phase, on the desired channel carrier

www.ingramcontent.com/pod-product-compliance
Lightning Source LLC
Chambersburg PA
CBHW021429180326
41458CB00001B/187